D0612787

The Politics of the
Global Oil Industry

The Politics of the Global Oil Industry

An Introduction

Toyin Falola and Ann Genova

PRAEGER

Westport, Connecticut
London

Library of Congress Cataloging-in-Publication Data

Falola, Toyin.
    The politics of the global oil industry : an introduction / Toyin Falola and Ann
Genova.
        p. cm.
    Includes bibliographical references and index.
    ISBN 0–275–98400–1 (alk. paper)
    1. Petroleum industry and trade—Political aspects. I. Genova, Ann. II. Title.
      HD9560.5.F34 2005
      382'.456655—dc22   2005013518

British Library Cataloguing in Publication Data is available.

Library of Congress Catalog Card Number: 2005013518
ISBN: 0–275–98400–1

First published in 2005

Praeger Publishers, 88 Post Road West, Westport, CT 06881
An imprint of Greenwood Publishing Group, Inc.
www.praeger.com

Printed in the United States of America

The paper used in this book complies with the
Permanent Paper Standard issued by the National
Information Standards Organization (Z39.48–1984).

10 9 8 7 6 5 4 3 2 1

To Jacquelynne McLellan and Lawrence Moore

Contents

Illustrations

FIGURES

MAPS

Preface

Since its discovery, petroleum has become the most lucrative and important industry in the world. This so-called black gold is considered one of the most important global resources. The oil industry transformed within just a few decades from a local to a global business, bringing into contact diverse peoples and cultures.

The writing of this book grew out of the need for a synthesis of international politics and oil, which affects everyone but is understood by few. Using a multidisciplinary approach, *The Politics of the Global Oil Industry* serves as an introduction to the oil industry in its international context. It captures the diversity and complexity of the global oil industry. It would be unrealistic to try to discuss all the world's oil-producing countries; instead, we have highlighted and explored the significant elements of the world's petroleum industry broadly. To make the book accessible to those unfamiliar with many of the oil-producing countries, we have limited our examples to seven major producers (Iraq, Mexico, Nigeria, Norway, Russia, Saudi Arabia, and Venezuela), while indicating many of the challenges that all oil-producing countries face. We hope this book will be useful to students as well as general readers who wish to become acquainted with the principal characters, events, and trends within the international oil industry.

This project would not have been possible without the support of our colleagues, family, and friends. We extend a special thanks to Jacquelynne McLellan for her invaluable comments on various drafts and many helpful suggestions. We would also like to thank those who shared in our enthusiasm for the project: Lawrence Moore and Akin Alao. This project would not have been possible without the help of our knowledgeable faculty and staff at the University of Texas, and we thank them for their assistance and support in our project. We thank the publishers for this opportunity, especially our editor, Hilary Claggett, who has been a delight to work with. For those moments of support, we wish to thank our families and friends: Ademola Babalola, Jonathan Brown, Peggy Brunache, Tyler Fleming, Matthew Heaton, Sander Kedich, Brandon Marsh, Jennifer Moore and Andrew Moore, Fehintola Mosadomi, Anne Turnbull, and Anne Webb.

Toyin Falola and Ann Genova
University of Texas at Austin

Abbreviations

AD	Acción Democrática
AGIP	Azienda Generale di Petrolio
AOC	Arabian Oil Company
API	American Petroleum Institute
Aramco	Arabian American Oil Company
Bcf	billion cubic feet
BP	British Petroleum
bpd	barrels per day
Caltex	California Texas Oil Company
CASOC	California Arabian Standard Oil Company
CFP	Compagnie Française des Pétroles
CPA	Coalition Provisional Authority
CVP	Corporación Venezolana del Petróleo
EEA	European Economic Area
EEZ	exclusive economic zone
ENI	Ente Nazionale Idrocarburi
ERA	Environmental Rights Action
ERI	EarthRights International
GATT	General Agreement on Tariffs and Trade

IEA	International Energy Agency
IMF	International Monetary Fund
INOC	Iraqi National Oil Company
IPC	Iraq Petroleum Company
IPE	International Petroleum Exchange
LNG	liquefied natural gas
MOSOP	Movement for the Survival of the Ogoni People
NAFTA	North American Free Trade Agreement
NNPC	Nigerian National Petroleum Corporation
NUPENG	National Union of Petroleum and Natural Gas Workers
NYMEX	New York Mercantile Exchange
OAPEC	Organization of Arab Petroleum Exporting Countries
OECD	Organization for Economic Cooperation and Development
OMPADEC	Oil Mineral Producing Areas Development Commission
OPEC	Organization of the Petroleum Exporting Countries
PDVSA	Petróleos de Venezuela SA
PEMEX	Petróleos Mexicanos
PENGASSAN	Petroleum and Natural Gas Senior Staff Association of Nigeria
SDFI	State Direct Financial Interest
SOC	Southern Oil Company
SOCAL	Standard Oil of California
SONJ	Standard Oil of New Jersey
Stanvac	Standard Vacuum Oil Company
Tcf	trillion cubic feet
TPC	Turkish Petroleum Company
UAE	United Arab Emirates
UN	United Nations
UNCLOS	United Nations Convention on the Law of the Sea
UNSC	United Nations Security Council
WB	World Bank Group
WTO	World Trade Organization

Part I

THE PLAYERS

Chapter 1

Essentials of Oil

Energy has become the currency of political and economic power, the determinant of the hierarchy of nations, a new marker, even, for success and material advancement. Access to energy has thus emerged as the over-riding imperative of the twenty-first century.

Paul Roberts, 2004[1]

What is petroleum? Why is it valuable? What is the connection between oil and politics? These are the crucial questions we seek to answer. In this book, we examine the world oil industry and the effect it has on the economics and politics of leading petroleum-exporting countries. The book serves as an introductory work that highlights the important aspects of the petroleum industry, offering cogent information about the countries, companies, and people that shape the world of oil. As an introductory work, this volume provides brief explanations of organizations and events. Suggested titles that offer more specific discussions are at the end of each chapter.

Three important points emerge in this book that will be briefly introduced here. First, and most obvious, oil represents a highly complex international issue. Second, oil and politics are intricately connected. Third, regardless of location, history, or culture, oil has represented both a blessing and a curse for all oil-producing nations. These themes will emerge numerous times throughout the book.

The book introduces the connection between oil and politics on a national and an international level. The oil companies operating throughout the world and the international organizations that formed around the oil industry are discussed. After laying the groundwork, we reinforce our more generalized discussions of the politics of oil with specific examples.

Using seven countries as case studies (Iraq, Mexico, Nigeria, Norway, Russia, Saudi Arabia, and Venezuela), we highlight the opportunities and challenges these countries face because of their extensive oil reserves. While there are numerous oil-producing countries in the world, we chose to focus on the world's largest oil-producing and exporting countries, particularly those that export petroleum products to the United States (see Figure 1.1). In many ways these countries acted as pioneers, taking the first step in a beneficial or harmful direction. Also, many of them are considered *developing countries* because they have recently become independent, they have economic problems that cause the standard of living to be low, and their economies are often dependent on external markets. For these countries, economic and political stability in the face of a lucrative yet volatile industry such as petroleum has posed a real challenge.

To understand the events that have unfolded in each of the chosen oil-producing countries, it is important to answer the three basic questions asked at the beginning of this chapter. First, what is petroleum? Many geologists believe that petroleum, or oil (the terms are used interchangeably), comes from the breakdown of plants and animals by some unknown process. Others, however, think that living things had nothing to do with its formation. Regardless of its origin, the mining of this so-called black gold has become the most lucrative and important industry in the world.

Oil appears to form below the ground through tiny openings, or pores, in rock that are visible through a magnifying glass. The rock formations from which oil is extracted are referred to as *oil reservoirs*. This term is often misunderstood, because an oil reservoir is not a large, cavelike space filled with oil. Instead, the rock formation acts like a sponge holding the oil and gas. Due to tectonic force and pressure, petroleum moves, usually upward toward the surface. By migrating, petroleum can accumulate in large amounts, particularly when oil and gas become trapped by rock folding or faulting. These traps range in size and shape. If the migration of petroleum is not blocked by geological formations, it will continue to move upward until it seeps at the surface.[2]

Second, why is petroleum so valuable? Petroleum is considered one of the most important resources in the world because people have used it for various reasons for more than 2,000 years, and new uses are constantly being discovered. Also, petroleum has served as a prime commodity that has linked the world together into the complex global market we see today.

Figure 1.1
World Crude Oil Production, 2003

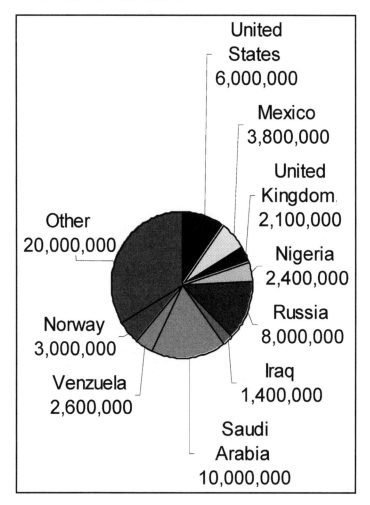

United
States
6,000,000

Mexico
3,800,000

United
Kingdom
2,100,000

Nigeria
2,400,000

Russia
8,000,000

Iraq
1,400,000

Saudi
Arabia
10,000,000

Venzuela
2,600,000

Norway
3,000,000

Other
20,000,000

Source: OPEC Annual Statistical Bulletin 2003 (Vienna: Organization of
the Petroleum Exporting Countries, 2004), 14; www.opec.org/library/
Annual%20statistical%20Bulletin/asb2003.htm.

What started as a regionally traded good and a gift brought to royalty by
explorers has today become a multibillion-dollar industry. The trade has
brought far corners of the world into contact with one another and has
turned oil into a politically strategic and economically lucrative commodity.

Although its "discovery" is generally recognized to have happened in the 1800s, oil was hardly unfamiliar to societies throughout the world before that time. The oil that seeps to the surface is what people in ancient times first discovered. It is usually found in the form of bitumen, which is a solid or viscous hydrocarbon. Today it is often processed into asphalt and is used primarily for construction of roads and for roofing materials. Around 3,000 B.C.E., ancient peoples of the present-day Middle East traded bitumen to be used as building mortar, medicine, and lighting fuel. Starting in 900 B.C.E., the Chinese began using natural gas from wells. The Aztecs of present-day Mexico used bitumen as medicine and mixed it with other ingredients to make dyes. They also used it on the roofs of their homes and on canoes as waterproofing material. In the 1500s the Spanish began to ship bitumen back to Spain from South America.[3]

Petroleum products caught on as widely traded commodities across the world in the late 1800s, when kerosene emerged as a major source of lighting fuel. Petroleum saw a second major boost in popularity with the invention of the automobile. Escalating demand for transportation fuel prompted oil companies to devote more crude oil to the development of gasoline than to kerosene.

Today, oil is understood as the most flexible fuel resource available. It can be used as an efficient source for heat, transportation fuel, and electricity. Beyond serving as a source of fuel, it is in almost everything we use in our daily activities. In addition to the most obvious examples of automobiles and gas heaters, petroleum products are in numerous other household items. They are in the clothing we wear, such as sweaters and shoes, and in any product that is made of plastic or rubber.

Finally, what is the connection between oil and politics? Oil has become an essential part of our everyday lives; there is hardly a moment when we are not taking advantage of the by-products of oil. With its wide array of uses, its popularity as a source of fuel, and the unique advantages it offers societies everywhere in the world, petroleum quickly became a major commodity traded and used in building nations economically and politically. Throughout history, petroleum has aided nations during military engagements by giving the nations that have it a strategic advantage over the have-nots. Economically, countries that had oil to trade set themselves apart as they became wealthier nations. Since the late 1800s, the international demand for petroleum products has increased so that oil-producing countries have found themselves able to achieve international recognition for their much-valued commodity.

FROM THE FIELD TO THE CONSUMER

To understand the politics of oil, it is critical to examine the phases of production. Many political conflicts emerge from various phases, so it is important to know how the oil industry works. Regardless of which company or country controls the oil, the process is the same. There are four major stages or phases to the oil industry: production (including exploration and extraction), refining, transporting, and marketing. There is a slight difference between crude oil production and natural gas production in the way the products are refined and transported.

In any discussion of the production process, a few terms and facts are useful to know. Two terms are generally used to differentiate the stages in the petroleum process: the upstream and the downstream. The *upstream* refers to the exploration, drilling, and production of petroleum; the *downstream* refers to the transport, refining, and petrochemical processes that get the oil to the consumer. In regard to the upstream, the term *crude oil* comes up frequently. *Crude oil* is the substance that oil workers pump up from under the ground. It is of little practical use in its natural state because it may contain impurities such as sulfur, oxygen, and metals, which must be removed before the final product can be delivered to the consumer. The two principle elements that make up crude oil are hydrogen and carbon, which is why the term *hydrocarbon* is often used in references to crude oil. Regardless of whether people are talking about crude oil or finished products that are ready to sell, you will often hear the more general terms *petroleum* or *oil* being used. When crude oil is extracted from the ground, it is said to have come from an *oil field*, which is the geographic area under which an oil or gas reservoir lies.

Oil fields differ not only in their geologic design, but also in the types of oil they contain. By geographic luck, some countries produce more valuable oil than others. There is a range of descriptive words that are used regarding crude oil that comes straight from the well. For the viscosity and weight, *heavy, medium*, and *light* are used. To determine the weight of the oil, companies refer to the API scale (an American Petroleum Institute [API] index). This scale is used internationally to classify crude oil and is written in degrees. Commercial-quality oil generally ranges from 20° to 50° API. The higher the index number, the lighter the crude. For example, a large percentage of Mexico's oil, traded as Maya-22, has a classification of 22° API, making it heavy. Saudi Arabia's Hawtah onshore field produces oil that has a classification of 45°–50° API, making this oil, known as Arab Super Light, the lightest in the world.

For the amount of sulfur and other impurities that must be removed from crude oil, *sour* and *sweet* are used. Sour crude is high in sulfur, while sweet crude is low in sulfur and requires less processing. The measurement for sulfur is fairly straightforward in that it is measured by the quantity present and is written as a percentage. In general, the amount of sulfur present ranges from 0.1 percent to 1.0 percent. Sweet crude is said to have 0.5 percent or less; sour crude has 0.6 percent or more. The Brent Blend from the North Sea has 0.37 percent sulfur, making it sweet. The market value of the oil is based partially on how much refining it requires and the level of impurities it contains. The heavier and more sour it is, the more time and money must be spent on extracting and refining the impurities out of it.

In regard to the downstream, the measurement terms come up frequently and can be a bit confusing. There are two major terms for how petroleum is measured: *barrels* (bbl) and *metric tonnes* (mt). Barrels are a volumetric unit of measurement that refers to large metal cylindrical containers that are in one internationally accepted size: 42 gallons (U.S.). Metric tonnes are a standard form used by the British for measuring mass. One mt of crude oil equals 7.33 bbl. However, since tonnes take weight into account, the type of oil is an important factor. For example, 8 bbl of Arab Light from Saudi Arabia are equal to 1 mt, while 7.5 bbl of Brent from the North Sea are roughly equal to 1 mt of crude oil. When the price of crude oil is discussed on the international market, however, barrels are always used.

Exploration and Production

Exploration and production of oil have never been a predictable or exact science. Much of the industry has been built on trial and error. A petroleum engineer described it best when he wrote, "Exploring for oil is an inherently discouraging activity." Even after numerous geological tests, only one out of ten test wells contains oil and only one out of a hundred successful test wells holds a commercial amount of oil. Once a major field has been discovered, it generally takes several years to develop the first shipment of crude oil.[4]

The exploration and initial production process require a great deal of time and capital because the terrain, the volatile nature of petroleum, and the political situation of the host country often pose challenges. Oil explorers hike through swamps, jungles, mountains, and deserts looking for possible oil wells. Once a well is found, roads must be built to move in equipment, and the power to run the equipment must be generated, all of which are costly. Aside from the cost, the discovery of a new well always holds technological challenges. In both onshore and offshore drilling, to avoid accidents and waste of valuable oil, oil workers must determine the amount of pressure in the well before commercial extraction can begin. When oil explorers first began to drill oil in the 1800s, they could not control the amount of pressure in the well. Numerous examples in the United States have been preserved in photographs of enormous gushers that occurred when oil had been struck. In the case of Spindletop, the first major field found on the gulf coast of Texas, the pressure in the oil well spewed crude oil and gas for nine days until the workers got it under control. On the other hand, wells that have been in operation for a while need additional pressure injected into them in order to extract the oil. Unfortunately for an oil-rich country, the discovery of oil does not guarantee that the money to develop the field is available. In the case of Russia, many new oil fields have been discovered, but the investment needed is heavy. The government is left to seek help from foreign companies, which generally find the current tax and legal situation in the country unfavorable. Nonetheless, oil companies are inherently risk takers, and the value of crude oil is often worth the uncertainty.

Crude oil is produced, or extracted, from two general locations: onshore and offshore. Most oil is extracted from onshore oil fields, while an increasing amount of oil is being produced from offshore sites. The term *offshore drilling* refers to extraction of oil from any depth in the ocean. In the past, oil companies had the technology to drill at a depth of only a couple of hundred feet; however, companies are now drilling in much deeper water, in depths of over 2,000 feet at distances hundreds of miles from the coast. Many of the countries that started their industry with onshore drilling are located on the coast and have simply shifted their exploration offshore. Unfortunately for landlocked countries such as Bolivia or those with relatively small coastlines such as Cameroon, offshore drilling does not hold a great deal of promise.

Natural gas may be produced from a gas reservoir, or it may be produced from a formation that produces both crude oil and gas. Natural gas that comes from a field without crude oil is known as *non-associated*. In Russia, for example, there are several natural gas fields in eastern Siberia. Most of the natural gas collected comes from crude oil production, which is known as *associated gas*. Only recently have companies collected and used gas while drilling for crude oil. Oil companies usually *flared* the gas off, which means they used a large flame to burn off the gas. Because of the environmental pollution gas flaring causes and the increasing value of natural gas, oil companies and governments together have proposed the elimination of gas flaring in the future. Countries such as Norway have taken a strong initiative in cutting the amount of gas flared. An international goal of expanding the natural gas market as well as the industry throughout the world has been proposed by the World Bank's Global Gas Flaring Reduction (GGFR) partnership.

Countries and companies discuss the size and value of the crude oil and natural gas fields in terms of their *reserves*. This term means that a field (which may include several wells) is estimated to have an estimated amount in barrels for crude oil or cubic feet for natural gas. For example, Russia has the largest natural gas reserves in the world, which total almost 1,700 trillion cubic feet (Tcf), and Saudi Arabia has the largest crude oil reserves in the world, which total 260 billion bbl. The trouble with discussing reserves is that the method of measurement varies. Legally, companies (private or state-owned) cannot include reserves for which no money to drill has been allocated. Oil in the ground maybe too costly to extract and once extracted not valuable on the world market. This kind of field is not supposed to be included in the reserve calculations.

Not all companies, however, abide by this method of measurement and often inflate their amount of reserves. As with most of the oil industry, politics are involved. Countries and companies have both over the years been accused of exaggerating their amount of reserves. Countries feel compelled to falsify reserves so that they look economically more stable than they really are, since many oil-producing countries rely heavily on oil revenues. Companies exaggerate in order to make shareholders feel secure in their investment. Reserves indicate a company's ability to make money in the future. Concerns over declining oil reserves available for production and trade may scare shareholders into selling, thus weakening the wealth of the company. In 2004 U.S. lawyers began the potentially long process of taking Royal Dutch/Shell to court for misleading investors by overstating its reserves from 1999 to 2004. Royal Dutch/Shell claimed that it exaggerated its reserves in Nigeria as a way of protecting the country.[5] Nonetheless, companies, governments, and international organizations, for the most part, rely on reported levels of reserves. They use reserves to rank and analyze the health of an oil company or a country's petroleum industry.

Tanks

After it is extracted from the ground, the crude oil is moved from the well to a production tank. Here the crude oil or gas from several wells converges to be measured, tested, separated, and treated before being sent to a refinery or gas-processing system. From the production tank, the quality and quantity of the oil and gas produced is reported to the producer, royalty owner, and purchaser. Governmental regulatory and taxing agencies also assess the oil produced. From here the crude oil and gas go to one of three places: through a pipeline to a port for overseas shipping, to a refinery, or to a storage tank. The storage tanks are designed for the storage and handling of large amounts of crude oil and are usually much larger than the production tanks. The storage tank is one of several places where crude oil is held when demand is lower than production. Some oil-producing countries take the excess oil, which would normally be placed in storage tanks, and inject it back into the ground. Not only does this process utilize a great storage place for surplus crude oil; in addition, the pressure can be used to pump out more crude oil. Storage helps regulate a country's crude oil industry to prevent too much oil from hitting the international market at one time and depressing oil prices.

In the case of natural gas, the measuring and storage processes vary slightly from those for crude oil. Once the natural gas has been separated from the crude oil, it generally requires further processing to remove water. This process is referred to as *dehydrating*. Contaminants such as carbon dioxide and compounds of sulfur also have to be removed, a process referred to as *sweetening* the gas. After sweetening, the hydrocarbons in liquid form are removed from the gas and stored in a low-pressure tank. In general this process is done near the field. However, when it is economical to gather the gas from several wells to a central point, then a gas-processing plant is built. This plant is the equivalent of a refinery for crude oil.

Pipelines

Pipelines represent a crucial component of the oil industry because they are the fastest, most efficient way to transport oil from one place to another. They are vast networks of gathering, transporting, and distributing systems comprised of hundreds of thousands of miles of pipe. These large steel tubes move oil or gas not only within a country, but also across national borders. For example, there is a pipeline that begins in Russia and runs through Belarus, Poland, and Germany. Pipelines transport both petroleum and natural gas, running below and above ground. In places where there are severe weather changes or permafrost, pipelines are run above ground, with several pipes running next to each other. This configuration allows the pipelines space to withstand thermal expansion. The downside to this, for the oil companies, is that the pipelines are vulnerable to sabotage or siphoning in unstable countries. For offshore production, the pipelines are often laid in a seafloor trench so that the pipes are not disturbed by fishing or other marine activity.

From the well to the production tank, the gas and oil travel together through a pipeline. After separation, each travels through its own pipeline network. Crude oil is moved through the pipelines via pumping stations, while gas is moved through the pipelines by a system of compressor stations. Pipelines carry the crude oil or gas either to be processed, or straight to an offshore transport terminal to be shipped. In the case of offshore drilling, the oil and gas that is gathered at a central drilling platform is transported to onshore facilities via underwater pipes. In a few cases, the platform has the capability to load the gathered oil and gas directly onto a ship for transport.

Refining

The crude oil and gas that are not directly transported via pipelines to port terminals are piped to a *refinery*. The refinery is a facility in a producing or nonproducing country, usually located near a port to allow for easy transfer of the oil to and from oil tankers. The refinery process involves transforming crude oil into usable products such as gasoline, kerosene, and aviation fuel, in quantities controlled by the refinery plant. The plant used to "refine" natural gas is not the same as a refinery. The removal of water, impurities, and excess liquids from natural gas is done either at the field or at a central *gas-processing plant*. After refining, most refinery products are ready to be sold to consumers. The remainder goes to petrochemical plants, where it is changed into other products such as rubber, textiles, and plastics. *Petrochemicals* consist of all the chemicals that are produced from partially refined oil and natural gas.

Oil refining started in the early nineteenth century with a patent made for the manufacture of "oil gas."[6] Since then, significant advances have been made in refineries' ability to refine and manufacture a variety of petroleum products in one location. Today's refineries require a great deal of financial investment and well-trained workers to operate the highly technical equipment.

Refineries vary in their capacity: a small refinery undertakes only a few processes, and a large refinery engages in the treatment of crude oil into several different forms. A refinery's capacity is the amount of petroleum that flows through it per day or per year and is expressed in barrels or metric tonnes. In general, the smallest refinery that is economically feasible has a capacity of 4,000 to 10,000 barrels per day (bpd) and may focus only on fuel oils. The economic success of a refinery often depends on the purpose of the refinery and the size of the community it serves. Smaller refineries are not the preferred choice for companies and governments, because the smaller the refinery the higher the costs to operate. Large refineries include those with a capacity from about 100,000 bpd up into the several hundred thousands. For example, each of Mexico's refineries has a capacity of between 200,000 and 300,000 bpd, while Saudi Arabia's largest refines 400,000 bpd. A refinery plant serves not only as a processing station, but also as a holding place. Most refineries have large storage facilities that act as an economic pressure valve, allowing a refinery to respond to dramatic price and demand fluctuations. For a major oil-producing country such as Saudi Arabia, these facilities act as a way of conserving the country's valuable resource.

Oil producing and non-oil-producing countries alike want to own and operate refineries within their countries. Through an oil refinery, a non-oil-producing country can engage in the world's oil industry. Also, in the long run it is more economically advantageous for a non-oil-producing country to refine crude oil than to import already refined petroleum products.

For both producing and nonproducing countries, several benefits come with owning a refinery. For a country that has *soft currency*, or a form of money that is not widely used or traded, owning a refinery is important because it reduces the amount of *hard currency*, such as U.S. dollars or euros, leaving the country. This is because finished petroleum products cost more than raw crude oil and because a local refinery can pay its workers in the local currency. Second, in addition to allowing greater control of a country's currency flows, refinery ownership allows a measure of control over what crude oil the country imports, how much the country processes, and who distributes the refined products to what locations. A country with a refinery can ensure that the domestic demand is met and then store the surplus for the future.

Third, for both producing and nonproducing countries, building and operating a refinery represent a positive national development project. Owning a large and crucial industrial complex such as a refinery provides a great deal of prestige, particularly to a developing country. A refinery represents a giant step toward economic growth and modernization. A head of state interested in establishing equitable development opportunities within his or her country can place the refinery strategically anywhere in the nation's boundaries. While it has been the trend to build refineries near ports, it is not an absolute requirement. Any easily accessible large plot of land in an area lacking any other economic opportunities serves as a reasonable possibility.

The layout and organization of a refinery allow a country to make a modest initial investment in a small refinery and expand as needed. A refinery can also be modernized and expanded to meet a variety of demands. Owning a refinery offers a country a step toward petrochemical production. Because petrochemicals have become such a necessary part of life, it makes sense to consider building a petrochemical complex as a future addition to a refinery.

Transporting

When petroleum products leave the refinery, they are transported by pipeline to a distribution center for regional consumption, or to a port for shipment overseas. For shipping petroleum, a specific type of ship called a *tanker* is used. Large tankers are used to carry oil across oceans and seas, while small tankers are used in coastal areas or along rivers. Crude oil tankers range in size from a supertanker, which has a capacity of about 3 million bbl, to a small tanker, which can carry about 150,000 bbl. Tankers also carry gas in liquid form. For the purpose of transport and processing, natural gas is liquefied. This requires that the holding tank be pressurized and refrigerated. The tanker is designed to withstand high pressure and low temperature in order to keep the gases in a liquid state.

While tankers are the most efficient way of transporting oil, they have been a cause of major environmental damage. In March 1989, Exxon's *Valdez* tanker hit a reef near Prince William Sound, Alaska, and spilled almost 300,000 bbl of crude oil. In November 2002, the independently owned *Prestige*, a 26-year-old tanker carrying 11,000 bbl, split in half, spilling oil off the coast of Spain. Since its sinking, thousands of barrels have continued to wash up on shore.

Another form of ship, the *barge*, is used for transporting a variety of petroleum products. It transports not only crude oil, like the tankers, but also petrochemicals, bitumen, and industrial fuel. A barge is a flat-decked vessel that lacks the power to move by itself. Instead, barges require tugboats and the water current to move them. Barges shuttle these petroleum products back and forth through a network of rivers, lakes, and channels.

The tankers and barges usually collect crude oil or natural liquefied gas from a port city that has an oil terminal. From the terminal, products are shipped to plants all over the world that will refine, distribute, and market the petroleum products for consumers, businesses, and airports. At the terminals, there are often storage areas, which provide a country with an opportunity to control its petroleum supply. Before a ship departs, final inspections are made regarding the quality of the crude oil or petroleum product.

Buying and Selling

Oil is the most important and most politicized commodity traded on the international market. While the price of oil is ultimately determined by basic supply and demand pressures, international politics plays a large part as well. Prices in the short and medium term reflect the political and economic situations of the world market—for the oil producers and the non–oil producers alike. Strikes, boycotts, embargos, environmental problems, and the financial health of an oil company all influence the market.

Petroleum is sold by the barrel and priced in U.S. dollars. All petroleum records and reports are given in English, making it the language of the petroleum market. This is because oil companies from the United States and Britain, for the most part, launched the international oil market and controlled the buying and selling of petroleum products.

From its beginning, the world oil market has been prone to wide price swings. As much as oil companies tried to create stability, the nature of oil exploration defied planning and scheduling. Production throughout the world lacked regulation. Often the market went from a shortage to a glut in a dizzyingly short period of time, though efforts to control the prices on the oil market have been made over the years. Who determines the pricing of oil has changed since the early years of the world market.

Three major shifts in who controlled the pricing of oil have taken place. From the late 1800s until the formation of the Organization of the Petroleum Exporting Countries (OPEC) in 1960, oil prices were published by the major oil companies. Eight interconnected oil companies, Exxon, Mobil, Standard Oil of California (SOCAL), Texaco, Gulf, British Petroleum (BP), Royal Dutch/Shell, and Compagnie Française des Pétroles (CFP), functioned in the crucial Middle Eastern oil fields through a complex web of joint ventures. By the 1950s these companies controlled nearly all aspects of the oil industry in almost all the fields outside North America, the Soviet Union, and China. Since the emergence of OPEC, this power dynamic has shifted. OPEC formed to take control from the major oil companies and to place it in the hands of the oil-producing countries. In its early years, the organization exerted a considerable amount of control over the world's oil. The members determined the prices by controlling the amount of oil produced each year. However, once their objective for forming had been met and non-OPEC oil-producing countries began to market their oil competitively, OPEC's control dissipated.[7] The 1990s marked a third shift in who controls the pricing. Today, pricing is largely determined on the trading floor, with thousands of traders and brokers playing a significant role in determining price. Most oil and natural gas is bought and sold, or traded, on the largest markets—the New York Mercantile Exchange (NYMEX), based in the United States, and the International Petroleum Exchange (IPE), based in London.

On the trading floor, there are several methods of buying and selling petroleum products. They include the traditional contract market, the spot market, and the futures market. Major oil companies shipping their petroleum prefer to have the majority of their oil sold in long-term contracts, with a portion designated for short-term, or one-time, contracts. The *traditional contract market* is the oldest method of trading oil in which major oil companies engage in a binding contract that lasts for a specified length of time. The price is fixed at the time of the sale and the contract length is usually several months to a year. The most common trading done, however, is on the *spot market*. In spot trading, a petroleum product is purchased "on the spot" at the current market price. Interested traders take part in these markets all over the world. Generally, brokers specialize in the trading of one form of petroleum. In the *futures contract market*, the buyer agrees to deliver or receive a set amount of oil (1,000-barrel minimum) in the future (usually in the following month) based on the present market price.

Thus far, we have addressed primarily the trading of crude oil or petroleum products. Natural gas is also traded, but not to the same extent. Its arrival as an internationally traded product is relatively recent. It appeared on the trading floor no earlier than the 1960s. Even today, the trading of natural gas is not truly an international phenomenon. It is not shipped to virtually every country in the world as oil is. Instead, natural gas is bought and sold within regions and is transported primarily via pipelines. Gas produced in Mexico, for example, goes to the United States, while gas produced in North Africa goes to Western Europe.

After natural gas, crude oil, and petroleum products are bought, sold, and traded, they are distributed and marketed. Basically, petroleum products such as gasoline are sold in two ways: by a vertically integrated oil company or by an independent marketer. A *vertically integrated oil company* is a major oil company that is involved in all aspects of the oil industry, from production to marketing. These companies typically sell their petroleum products in their own service stations. An *independent marketer*, however, operates only in distributing and marketing. Independent marketers often buy the surplus products from multinational companies or independent refiners. In some cases, the independent marketer transports the petroleum products in its own trucks from the refiner to the service station or retail outlet. They may also maintain storage facilities. An independent company can hold either a long-term or short-term contract with a refiner.

The marketing of petrochemicals differs slightly from the marketing of common petroleum products. The demand for petrochemical products fluctuates just as do the other segments of the oil industry. The difference is that petrochemicals are for the most part not end products, but act as intermediates for the manufacture of other petrochemicals. These products are sold to manufacturers and made into household products. Because petrochemicals pass through so many hands, many vertically integrated oil companies have stayed out of petrochemical marketing. Instead, they sell their refined products to chemical companies or manufacturers who turn the petrochemicals into consumer goods.[8]

The production process of the oil industry is highly complex, requiring the best facilities, the most knowledgeable oil workers, and the collective effort of the companies and countries involved. Without these components, the industry and countries dependent on it suffer. For most oil-producing countries, the challenge has been striking a balance between political stability and economic expansion within the petroleum industry. The multifaceted oil industry, with its processes taking place all over the world and its complicated network of contracts and agreements, only adds to the challenge. Many oil-producing countries have confronted the situation and emerged either triumphant or destroyed. Oil has the power to pull a country into international politics and push it to take political action based on the prospect of accumulating more oil.

THE POWER OF OIL

Oil has the power not only to catapult a country into international politics, but also to entice an oil-rich country to pursue more power. Many oil-producing countries have acquired this privileged status within the past seventy years. Some countries have been in control of their own industry for only the past forty years. As discussed earlier, it takes at least ten years from the discovery of oil for a country to see a return on its investment. For many countries, noticeable differences occurred as they became major oil-producing countries. These countries witnessed an increased number of visiting dignitaries and an often enhanced position within the United Nations.

Along with the prestige comes a growing sense of obligation. Countries may feel compelled to share their production with the world and their wealth with their nonproducing neighbors. At the same time, they feel required to actively engage in world politics in order to protect their production, maintain their access to markets, and have a say in pricing. Oil-producing countries have two main goals: to expand their oil industry and to collect as much revenue as possible from it. While these goals appear fairly simple, oil-producing countries do not exist in a vacuum. A history of economic and political instability within a country and poor regional relations may delay progress. In many cases, the presence of an oil industry has aggravated already troubled situations such as poverty and corruption. Oil influences almost every aspect of a country. A country's current situation and the future of its petroleum industry affect internal development plans, foreign policy, environmental and civil rights laws, and government-corporate relations.

The often changing political and economic situation within an oil-producing country adds to the already fluctuating nature of the world oil market. A producing country, such as Norway, that does not face problems of political corruption, widespread poverty, or regional conflict, is still vulnerable to any global changes that occur with regard to oil. Fluctuations within the international petroleum industry often depend on the discovery of a new oil field, the exhaustion of an old oil field, war, and political change or instability within an oil-producing country. For instance, because of the Gulf War, Iraq went from producing 2.8 million bpd in 1989 to producing 282,000 bpd in 1991. Within the past few years, political problems in Venezuela, Iraq, and Nigeria have pushed Saudi Arabia to up its production levels to more than 8 million bpd.

Trading relationships between an oil-importing country and an oil-producing country often change based on politics as well. Within the past couple of years, the United States has increasingly imported petroleum products from countries within the former Soviet Union. Since the majority of major oil companies are from the United States, international interest in oil-producing countries shifts along with U.S. trade patterns. The Middle East has been the focal point of the oil world for many years, but this trend may shift. The new fields in the Caspian Sea countries are the current focus of major oil companies.

The impact of a disruption in production varies from country to country. This is because not all importing and exporting countries are directly connected to one another. Who imports oil from where depends on the complex arrangements among the producing country, the oil companies operating there, and the importing country, as well as physical proximity. For this reason, a major oil producer's oil may not be exported to every importing country. For example, Saudi Arabia is the world's largest oil producer and exporter. The United States imports the bulk of its crude oil from Saudi Arabia; however, Saudi Arabia's crude does not generally go to South American countries. China exports its petroleum products to the United States and countries in the Pacific, but not to Europe.

Although an oil-producing country may produce millions of barrels every day, this does not mean the country is self-sufficient. Most oil-producing countries designate certain oil wells and refineries to satisfy their domestic market. However, some oil-producing countries either do not make this arrangement or do not reserve enough and are forced to import oil until the problem is remedied. Some countries seek only to produce what their country needs without being overly concerned with exporting their petroleum. On the other hand, some oil-producing countries are extremely preoccupied with exporting as much oil as they can produce and do not retain enough for their own citizens. In many cases, being an oil producer does not guarantee that an energy crisis within the country will never occur.

Some oil-producing countries find that their people cannot afford to pay the market value for petroleum products. This is particularly the case in a country that produces highly valued crude oil. In the case of Russia, local oil companies prefer to export the oil instead of selling within the country because they can make more money per barrel on the international market. The solution for some countries has been to subsidize the oil they produce for the domestic market. In practical terms this means that if a petroleum product costs $5 on the international market, for example, the consumer in the producing country will pay $3 and the government will pay $2. Also, some countries, despite changes in the market price for their products, will continue to charge the same amount year after year. This means that the government's subsidy percentage continues to increase, while there is no change in price for the consumer. This is when subsidizing becomes a serious problem for an oil-producing country.

Oil-producing countries use their oil to build regional and international alliances in several ways. Aid, military strength, and cross-country pipelines are just a few examples. The aid takes many forms, including cash for humanitarian reasons, and emergency relief aid in the form of clothing, food, medical supplies, and technical assistance. OPEC members also give aid to their neighbors indirectly by contributing to the OPEC Fund, which provides aid to non-Western developing countries.

Oil-producing countries also use their oil to satisfy their own nationalist agendas as well as supporting other nationalist movements throughout the world. The most notable example of using oil as a political weapon to satisfy national interests occurred during the 1960s with the formation of OPEC. Five major oil-producing countries (Iran, Iraq, Kuwait, Saudi Arabia, and Venezuela) primarily in the Middle East decided to take direct control of their oil industry and pricing away from the major oil companies (see Chapter 4 for a more in-depth discussion). With nationalistic fervor, OPEC managed to achieve its goal. Each member agreed to own at least 50 percent of all oil operations in its respective country. Furthermore, oil-producing countries have provided assistance by supplying liberation movements with military support. For example, during the 1970s, members of the Arab Summit Conference assumed a leadership role in anticolonial and antiapartheid struggles in Zimbabwe and South Africa by providing financial assistance to the liberation movements.

Embargos have been used by oil-producing countries, and this political tactic has also been used against them. Throughout history, numerous organizations and political strategists have called on governments to use oil to seek political change. Within the past ten years, calls have been made for an international embargo on Iraq and Nigeria in order to place pressure on the military regimes. In the case of Iraq, U.S. and UN economic sanctions had been in place since 1990. The United States imposed a complete trade embargo, making importation to or exportation from Iraq of any goods or services illegal. In May 2003, both sanctions were lifted in order to provide aid to Iraq's war-weary people after the invasion. In the case of Nigeria, the embargo stemmed from the growing concern within and outside Nigeria over the violation of human rights in the Niger Delta region during the mid-1990s. The sanctions had been imposed on Nigeria by the United States, Canada, and the European Union, which meant that business activities between Nigeria and these entities were significantly reduced.

As in the case of Iraq during the Gulf War, oil also causes conflict. Border disputes, outright oil-hungry invasions of neighboring countries, and mutual claims to offshore territories have all plagued the world of oil. Border disputes become intense, especially when known reserves are involved.

In the 1930s the British established the Saudi-Kuwaiti Neutral Zone due to political conflict. After the discovery of oil in the late 1930s, both countries agreed to share the oil wealth, each taking half. The Saudi-Kuwaiti Neutral Zone has flourished, with a major oil field that produces over 600,000 bbl of oil per day.

Mutual claims to offshore oil fields have also been a problem. International law allows a country to claim ownership from its coast to a distance of 200 nautical miles, approximately 230 distance miles. In the 1970s, a political clash between Nigeria and Cameroon was triggered by mutual claims to offshore oil reserves along the southern portion of the common border known as the Bakassi Peninsula, located in the Gulf of Guinea. An international court granted the peninsula to Cameroon in 2000. In part of the South China Sea, a promising petroleum province remains unexplored because the drilling rights are claimed by six different countries. In the case of Saudi Arabia and Bahrain, the solution to a mutual claim of offshore oil has been to produce it jointly. Furthermore, some of the profits made from this oilfield by Saudi Arabia are donated by Saudi's government as an attempt for regional peace back to Bahrain.

Oil-producing countries have also satisfied the demands of national interest and regional cooperation by building pipelines. Pipelines give landlocked oil-producing countries a way of transporting their oil to a port terminal and to their regional neighbors along the pipeline route. In the case of Venezuela, the mutual interests of Venezuela and Trinidad and Tobago in gas fields have introduced the idea of building a pipeline in order to extract the gas in a cooperative manner. Pipelines for natural gas, particularly, have been emerging within the last several years, because this method of transport is cost-effective for natural gas.

Oil does two things for an oil-producing country that conflict with each other. The economic and political strength of oil can turn a country into a wealthy country but at the same time instill an oil-producing government with a relentless drive to maximize profits and keep the oil flowing. This often creates severe problems within and outside the country. In other words, for oil-producing nations, oil has been both a blessing and a curse. In countries that did not establish political stability prior to developing an oil industry, achieving economic growth and stability within a country that is often plagued by corruption, poverty, and cultural tension can be very difficult.

But why is the prestige of being an oil-rich country so fraught with problems? To best understand why oil has this effect, we need to closely examine the development of these countries' oil industries. For this, we begin with a close examination of the oil companies that brought commercial production to these countries.

FURTHER READING

For a good introduction to the technical side of the global oil industry, see Kate Van Dyke's *Fundamentals of Petroleum* (Austin: University of Texas Press, 1997) and Robert Anderson's *Fundamentals of the Petroleum Industry* (Norman: University of Oklahoma Press, 1984). See also OPEC's *Basic Oil Industry Information* (Vienna: OPEC, 1983).

Chapter 2

The Oil Companies

I have ways of making money you know nothing of.
John D. Rockefeller, 1872[1]

In this chapter we explore the role of oil companies who not only managed the international oil industry but also dominated it for many years, and in some ways still do. Here we highlight the companies and their origins. Not all oil companies are the same in size, philosophy, and operations. This is most evident in their methods of interaction with one another and the oil-producing countries in which they operate. Oil companies over the years have learned how to balance competition and cooperation among one another while satisfying an oil-rich country's interest in being an active partner in a foreign company's operations. To do so, oil companies organized themselves in a number of ways.

The most common method is the basic *holding corporation*, examples of which are ExxonMobil and BP, which is an artificial body that possesses properties distinct from the companies that compose it. In the case of ExxonMobil, Mobil, Exxon, and their subsidiaries have some level of autonomy within ExxonMobil. Many times corporations such as ExxonMobil enter into long-lasting business relationships with another oil company. These *joint-stock companies* can include several oil companies (state-owned and independent alike) that receive shares for the capital they invest. In these joint-stock companies, each company actively takes part in the joint-stock company's operations but may sell its shares within the parameters of the agreement. When commercial production began in Iraq, oil companies from the United States were originally left out of the

Iraqi Petroleum Company, a joint-stock company operating in Iraq, until BP agreed to sell 20 percent of its shares to Exxon in 1922. A few years after joining, however, Exxon withdrew to pursue other production interests. In addition to how they are structured as businesses, oil companies also vary in ownership and size.

Three principle types of oil companies constitute the international oil industry: the major oil companies and their subsidiaries; the minor companies; and the state-owned companies. The major oil companies, such as ExxonMobil, BP, and Royal Dutch/Shell, are often vertically integrated, having thousands of employees running their operations all over the world. Their subsidiary companies acts as semi-independent parts of a larger parent oil company. For example, Esso serves as ExxonMobil's marketing company outside the United States. In *Fortune*'s Global 500, which ranks the world's most profitable companies, the major oil companies ranked within the top ten, with BP as the second most profitable, Exxon-Mobil third, and Royal Dutch/Shell fourth.[2]

The minor companies, such as ConocoPhillips, ChevronTexaco, and Ente Nazionale Idrocarburi (ENI), are similar to the majors but operate, overall, on a slightly smaller scale. Along with the majors, they rank among the largest companies in the world. TotalFinaElf is the world's fourth largest oil company, ChevronTexaco fifth, ConocoPhillips sixth, and Eni seventh. They often focus on marketing petroleum products, with partial ownership in a production company. The state-owned oil companies often do not operate outside of their country and are 50 to 100 percent owned by their government. Often these companies act as the sole producers and providers for their country, as does Mexico's PEMEX, or they act as a joint-stock partner with a multinational company, as in the case of Venezuela's PDVSA. These different forms of oil companies offer a glimpse into the complexity and uniqueness of the oil companies.

Although the structure of oil companies as well as the way they conduct business may vary, they have several general characteristics in common. First, the companies operate on an international level not by choice, but because their interest in petroleum takes them to the location of the resources, whether on the plains of Oklahoma or in the wetlands of Nigeria. Therefore, oil companies became multinational out of necessity, to accumulate profits. Second, oil companies must function as risk takers and express a willingness to invest heavily. These companies find that oil testing and production cost a great deal of money and yield unpredictable results. Finally, the companies need to be aware not only of government policies and regulations within the country of operation, but also of international laws pertaining to oil development and trade. For these reasons, oil companies are often quite large in comparison to other businesses and have a huge amount of capital available. These factors make oil companies unique within the world of international business.[3]

Much of the oil companies' history lies in their development through competition with one another. The history of these oil companies appears contradictory in that the same competition that drove each company to expand aggressively into the far corners of the world forced them into business agreements and alliances with one another. The result of this dichotomy has been the continual reconfiguration of the companies through mergers, joint ventures, and the creation of affiliated companies. Each company underwent a series of name changes reflecting its restructuring, and often this change is not reflected in all aspects of its operations. When a merger occurs, retail gas stations, for example, do not always reflect this change. Also, for marketing purposes, the oil companies sometimes prefer to maintain familiar logos to maintain customer loyalty to a product. For this reason, their history appears complex.

The simplest way to understand the oil companies is to examine them in their formative years, when they acted relatively independently in fierce competition with one another. To understand one of the world's major oil companies, ExxonMobil, the story of John D. Rockefeller and his Standard Oil Company provides an excellent starting point.

THE ROCKEFELLER STORY

In the middle of the nineteenth century, a man named John D. Rockefeller revolutionized the oil industry in the United States. Before the formation of his enormous oil company, Standard Oil, petroleum products were produced, transported, and sold by small independent companies operating on a regional level. In 1865, Rockefeller joined two other men in owning a refinery in Cleveland. After buying out the two other men, he saw that the refinery business could allow him to dominate the oil industry. He stayed out of oil production because he viewed it as a risky business venture. Rockefeller is often considered the father of *horizontal integration* in the United States because he bought out competitor after competitor in the refining industry. His edge over his competitors stemmed from the secret rebates he convinced the railroads to give him so that he could transport his petroleum products farther and for less money than other refiners. To his competitors, he offered two options: sell

Figure 2.1
Deconstruction of Standard Oil[4]

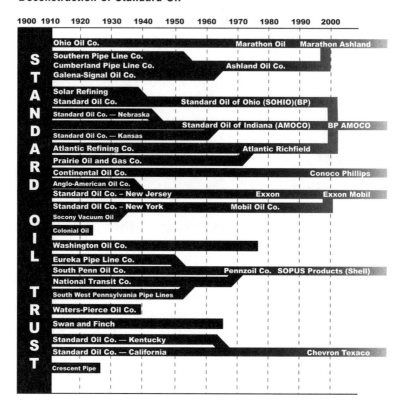

their companies to him or count the days until he drove them out of business. If they agreed to sell their companies, he promised to let them continue managing their businesses under the control of Standard Oil (of Ohio).

Until 1890, Rockefeller continued to expand his company across the United States by buying up more refineries and building a pipeline system, which brought in millions of dollars for the company. Standard Oil continued to expand beyond the eastern United States, establishing itself as the nation's dominant oil company. In 1882, all of Rockefeller's oil operations were merged into the Standard Oil Trust, which created a board of nine trustees, who collected profits from all of Standard Oil's activities and divided them among the company's stockholders. A large part of Rockefeller's success came from the clever organization of Standard Oil in a way that side-stepped the law preventing a company in one state from owning shares in a company in another state. By allowing the bought-out companies to appear independent, he did not break any laws. Of particular importance was the formation of Esso, which served as the distribution and sales subsidiary of the new Standard Oil in the United States, and, more frequently, overseas. The name came from the company name Eastern States Standard Oil Company, which was shortened and used to avoid the stigma associated with "Standard Oil" in 1932. Any concern over the legality of Standard Oil with the state and federal governments was dampened by Rockefeller's political friendships and skilled team of lawyers.

Aside from the legality of the trust was the national concern regarding Rockefeller's ability to control oil prices. Standard Oil had extended its control not only over its competitors, but also over oil transportation. Nearly every method of transport from the oil fields to the consumer was owned by Standard Oil, which allowed the company significant control over prices. The public's growing dependence on kerosene for lighting created consumer frustration and resulted in the public's appeal to state and federal governments to take action against Standard Oil.

In response to the public outcry, both the state and federal governments began to investigate Standard Oil's activities shortly after the formation of the trust. Ten years after its formation, a court decision in Ohio dissolved the Standard Oil Trust based on the Sherman Anti-Trust Act of 1890. Senator John Sherman's goal included the prevention of monopoly formation and the protection of independent businesses. In 1888, much to Rockefeller's luck, New Jersey passed a law that allowed companies registered within the state to hold shares of companies outside it. Rockefeller reorganized the Standard Oil Trust into Standard Oil of New Jersey, which acted as a holding company to the other operations outside the state. Essentially, not much changed except the name and location of the company. However, this move did not make the company less vulnerable to attacks from the federal government, which expressed a growing interest in breaking up corporate monopolies within the country. Public grievance forced the company into the federal court system and, in 1911, the Supreme Court demanded that the company break up into several independent companies. This radical decree returned each subsidiary back to the state where it operated, but they remained in the hands of Rockefeller's partners; Rockefeller had retired from the business in 1897. The federal court decision never fully disabled Standard Oil's strength. Out of the thirty-four companies that made up Standard Oil, three became major multinational corporations. Despite the government's attempts in 1911 to break up Standard Oil, the recent merger of Exxon and Mobil, both from

Standard Oil, show that its efforts did not last. Furthermore, the law did not prevent the companies from expanding and dominating the oil industry overseas (Figure 2.1).

THE MAJOR OIL COMPANIES

Prior to the ExxonMobil, BP, and ChevronTexaco mergers, the major oil companies were commonly referred to as the Seven Sisters, which included Exxon, Mobil, Chevron, Texaco, Gulf, British Petroleum, and Royal Dutch/Shell. This was because Exxon, Mobil, and Chevron came from the parent company Standard Oil and, like many real sisters, the seven major oil companies had a love-hate relationship. They competed with each other while at the same time they built alliances, particularly when threatened by outsiders.[5]

The companies within the United States developed in a slightly different way from their European competitors. Oil companies from the United States began in a protected environment. They produced oil within the country and did not face the risk of being undercut by imports. Thus, they developed a strong market base and accumulated wealth within the country before expanding outside. This was aided by the fact that the United States until World War I favored national protection by closing its doors to outside investors. The British and Dutch companies, on the other hand, had not discovered oil at home and were forced from the beginning to find oil supplies outside their countries. In the nineteenth century, trading companies from Europe had already begun searching for oil in the Middle East, Africa, and Asia. Trade for these companies meant long distances across the Pacific Ocean or Mediterranean Sea, while in the United States trade developed regionally and then expanded as the technology of pipelines and railroads developed. In other words, these European companies operated at an escalated risk level because they not only needed to locate oil in faraway places, but also had to negotiate getting the oil back home. Regardless of their formations, however, oil companies based in the United States and Europe quickly became competitors on an equal footing.

Public opinion toward the companies from the United States and Europe also differed. By the time the separated Standard Oil companies ventured overseas, the company and all its former components had been the focus of muckraking journalism and multiple court cases on the state and federal levels. For the most part, the public viewed Rockefeller and Standard Oil as ruthless monopolists who operated with a blatant disregard for consumers. In Europe this problem did not exist. There were no antitrust cases or frustrations over oil pricing that went to the judicial system. The government and the oil companies worked more closely together and understood their mutual need for one another. For this reason, these companies did not outwardly exude the same profit-driven attitude that existed within the United States. The oil companies, particularly in Britain, displayed a commitment to the empire.[6] This is not to say that the governments did not debate over their relationship with the oil companies. Oil was a highly valued commodity that, perhaps, they thought, should not be left solely to the oil companies. The United States government as well as those in Europe questioned whether the companies should be controlled, and if so to what extent. This concern stemmed from the fact that as these oil companies expanded their operations, their level of control over the industry increased. Within a few decades after World War II, these few major oil companies controlled nearly the entire international oil industry.

ExxonMobil

To complete the story of Standard Oil, we begin here by discussing the most significant companies to come out of the dissolution of the giant company. The largest component of the company, Standard Oil of New Jersey (SONJ) posed as the holding company for all of Standard Oil until 1911, when the U.S. Supreme Court called for the dissolution of the Standard Oil Trust. In 1954 SONJ moved to acquire shares of its successful affiliate company, Humble Oil and Refining Company. Five years later, SONJ had purchased all of Humble, more than doubling SONJ's assets. The company operated under the name Standard Oil of New Jersey until 1972, when it changed its name to Exxon Corporation. Like the other newly independent Standard Oil companies, the company found itself cut off from crude oil that had been produced by the other members. Realizing the problem, Exxon went overseas to carry out its exploration activities in Mexico, Venezuela, and Norway, to name just a few. By the early twentieth century, SONJ had become one of the largest oil-producing companies in the world.

At the same time, its neighbor, another former Standard Oil offspring, Standard Oil of New York, known as Socony, had formed in the late 1880s. Operating largely as a marketing company, it merged with Vacuum Oil Co., in 1931, to form Socony-Vacuum. The Vacuum Oil Company had also been part of Rockefeller's trust. Socony-Vacuum extended its marketing operations throughout Asia but lacked a reliable source of oil.

Realizing their need for each other, SONJ and Socony-Vacuum merged their interests in Asia in 1933. SONJ produced oil and refined it in Indonesia but had no marketing network, while Socony-Vacuum had the markets. The agreement formed a 50-50 joint venture called Standard-Vacuum Oil, commonly referred to as Stanvac, which was dissolved in 1962. By 1966, Socony-Vacuum became known as simply Mobil Oil Corp. Similarly, SONJ underwent a name change in 1972, when it became known as the Exxon Corporation. Outside the United States, however, it retained the name Esso Petroleum Company. In 1999 Exxon and Mobil merged into ExxonMobil. Today ExxonMobil and its affiliated companies operate in more than 200 countries. Two years later, another former Standard Oil component, Standard Oil of California (SOCAL), engaged in a monumental corporate merger.

Royal Dutch/Shell

Royal Dutch/Shell began in the 1800s as two separate companies, Royal Dutch and the Shell Transport and Trading Company. Henri Deterding formed Royal Dutch in the Netherlands as a company that received a royal charter from the Dutch government to explore Asia during the 1890s. Marcus Samuel began Marcus Samuel and Company in 1833 selling antiques and exotic seashells in Britain. In 1890, Samuel broadened its interests into the kerosene market developing in Russia. In 1891, Samuel signed a nine-year contract with the powerful Rothschild family to buy oil.[7] The Rothschilds were of German descent but had built their wealth in France. Through this arrangement Samuel began to transport kerosene from Russia to Asia. In 1897, he formed the Shell Transport and Trading Company. With its fleet, its markets, and its capital, Shell appeared much stronger than Royal Dutch. In order to solidify its position, Shell's owner, Marcus Samuel, made an extraordinary deal with Gulf Oil. The contract stated that Shell would buy almost half the company's production and take it to Europe. Unfortunately for Shell, the deal soured when Spindletop dried up in 1902 and Gulf could no longer meet Shell's demands. Having lost its supply of oil, Shell realized the value of working with Royal Dutch instead of competing, and so they merged into one company in 1907. Since the merger, Royal Dutch/Shell has developed into an astoundingly successful oil company with operations all over the world. During the second half of the twentieth century, Shell expanded its operations through discoveries of major oil fields in places such as the North Sea and Nigeria. In the late 1990s, Shell, Texaco, and Saudi Aramco agreed to combine their refining and marketing businesses in the southern and eastern parts of the United States. By 2004, Shell had converted almost 5,000 Texaco retail stations into Shell stations. With this rapid rate of expansion, Shell has become one of the world's leading suppliers of petroleum products.

While the formation of Royal Dutch/Shell saved Shell from the threat of financial ruin, it also alienated Shell from the British government because of its alliance with a Dutch company. Shell's activities, to the British government, no longer embodied British interests. A feared alliance between the Dutch and the Germans made the British government uneasy, particularly since the main oil supply for Shell came from Dutch territory in the Far East. Since the late 1800s, Samuel had campaigned for the Royal Navy to switch from coal to oil as a source of fuel, but the navy resisted. By the early twentieth century, any hope of Shell becoming the main supplier of oil for the Royal Navy had been dashed. When Winston Churchill set out to secure an oil supply at the onset of World War I, British Petroleum had become the company of choice. The British Petroleum Company fulfilled Britain's nationalistic interests perfectly. The British by no means withdrew all interest from Shell but, instead, kept them at a distance while they developed a very strong relationship with British Petroleum. Regardless, Shell established itself as one of the world's largest oil companies almost from the beginning and continues to expand today. Currently Royal Dutch/Shell is owned by shareholders and has operations in more than 145 countries.

British Petroleum

Around the same time Royal Dutch/Shell formed, British Petroleum made its way onto the international oil stage. Like all the major oil companies, it has several phases to its history. The formation of the company started with a British business-minded adventurer, William D'Arcy, who went in search of oil fields in Persia (present-day Iran) in 1901. His company in 1909 became the Anglo-Persian Oil Company, which controlled almost all of Persia's oil fields. In 1914 Winston Churchill, a member of the War Council of Great Britain at the time, bought half of the company for the British government in order to assure oil supplies for the navy and secure a significant share in oil concessions in the Middle East. In 1935, to reflect the name change of Persia to Iran, the Anglo-Persian Oil Company became the Anglo-Iranian Oil Company.

Until the 1960s, Anglo-Iranian did not have operations outside the Middle East. Iran remained the main center of its operations. In the 1950s, however, Anglo-Iranian lost its stronghold in Iran through a bitter conflict over the host country's interest in nationalization. To improve its image after the crisis in Iran, the company changed its name in 1954 to British Petroleum. A year later it restructured to become a holding company, and BP Trade Limited became a subsidiary that managed trading activities. In 1987 the British government sold its stock in the company and the company acquired the remaining stock of Standard Oil of Ohio as well as the British company Britoil. In 1998 Amoco, which was known as Standard Oil of Indiana before the dissolution of the trust, merged with British Petroleum. In 2001, British Petroleum changed its name to simply BP, stating that the "B" and "P" stand for Beyond Petroleum.

Today, BP, including its subsidiaries, operates in 100 countries all over the world. In 2003 BP became 50 percent owner of the newly formed Russian oil company TNK, which combined Russian petroleum assets of BP, Tyumen Oil Company, and Sidanko. After the company's formation, BP claimed that it surpassed Royal Dutch/Shell as the world's second largest privately owned oil producer. BP also holds a 25 percent stake in Slavneft, a Russian oil company, through its 50 percent TNK-BP holding. Since its formation, BP has played an integral role in the global oil industry.

These major oil companies all became powerful in the oil industry virtually from their beginnings. Each of them vertically integrated, which allowed them to control every aspect of the oil business. This goal of self-sufficiency, however, could not be attained without the aid of the minor oil companies.

THE MINOR COMPANIES

The minor companies, also known as the independent companies, are those without a global system of marketing, producing, transporting, or refining, yet they have played a significant role in the development of the international oil industry. These companies, totaling twenty to thirty, ventured into the international oil industry during the 1950s and 1960s and are primarily companies from the United States and Europe. What makes the categorization of these companies as "minor" ironic is that in any other industry, the wealth and size of these companies would make them the dominant companies. As "minor" oil companies they still rank within the top twenty largest companies in the world. Several of them originated from the Rockefeller trust, such as Standard Oil of California (SOCAL), Standard Oil of Indiana (Amoco) and Continental (Conoco).[8] The minor companies outside the United States include ENI and TotalFinaElf. Despite the wealth and economic foothold these companies have accumulated throughout the world, they are seen as struggling underdogs next to the corporate giants. This does not mean the major and minor companies operate in opposition, however.

The minor companies often fill domestic needs while the large companies take their production interests overseas. The major and minor companies over the years have formed joint-stock companies, merged, and recognized one another's positions within the global market. Here we will introduce a few of the minor companies, not only to provide examples, but also to illustrate their relationship to the major companies and one another over time.

ChevronTexaco

The ChevronTexaco Corporation, a conglomeration of Chevron, Texaco, and Caltex, formed in 2001, but each company has its own unique history dating back several decades. Chevron, known as SOCAL until 1977, began in 1879 with the discovery of oil north of Los Angeles. This discovery led to the formation of the Pacific Coast Oil Company. Rockefeller, who had already set up a distribution office in 1878 for his oil in San Francisco, used his usual style to push the Pacific Coast Oil Company into becoming part of Standard Oil in the 1880s. As part of the agreement, Pacific Coast retained its name until 1906, when it became Standard Oil of California. This company, as part of Standard Oil, lasted only five years before the Sherman Anti-Trust Act separated the company from its parent company in 1911. Continuing the Rockefeller tradition, SOCAL acquired several other regional oil companies, such as Murphy Oil Company (1913) and Pacific Oil Company (1926). After the opening of the Panama Canal in 1914, SOCAL gained greater access to the eastern United States' and Europe's markets. As Standard Oil of California, the company only operated under the Rockefellers for eleven years before the dissolution. Once free to pursue its own interests, SOCAL expanded its production operations. While SOCAL expanded during the early twentieth century in California, its future partner, Texaco, also had its share of success in Texas.

In 1901 two men (one from Pennsylvania and the other from New York) established the Texas Fuel Company, which opened its first office in Beaumont, Texas. In 1902, the Texas Fuel Company expanded into the

Texas Company, later to become the Texas Corporation in 1926 when the company moved its headquarters to Delaware. By 1904, the company had set up nationwide sales centers in order to compete with Standard Oil and, in fact, did quite well. In 1941, the company consolidated its subsidiary companies based throughout the United States and took the name Texas Company once again. In 1959, the company became Texaco Incorporated. From its formation, the company managed to maintain its competitive edge, particularly against the former Standard Oil companies.

While Texaco expanded its operations, SOCAL did the same. SOCAL expanded its exploration overseas to places such as the Philippines, Bahrain, Saudi Arabia, and Mexico. In 1936 the two companies teamed up in the Bahamas to form the California Texas Oil Company Limited, better known as Caltex, to exploit a potential market in the Middle East and later in Africa, Australia, Europe, and New Zealand. In 1946 Caltex moved its headquarters to Delaware and changed its name to the California Texas Corporation. Two more name changes occurred, and the company became Caltex Petroleum Corporation in 1968. SOCAL and Texaco remained independent aside from their joint company, Caltex, and each company expanded. In 1961 SOCAL merged with Standard Oil of Kentucky, which was yet another former Standard Oil offspring. In Western Europe, SOCAL and Texaco agreed to dissolve Caltex and split its operations between the two parent companies, SOCAL and Texaco. In order to manage its share of Caltex's operations in Europe, SOCAL formed Chevron Oil to handling the marketing operations, in 1967.

In 1984 SOCAL became Chevron Corporation to better incorporate its marketing wing. Chevron and Texaco merged into one company, ChevronTexaco, in 2001. Today ChevronTexaco operates in 180 countries. While Texaco represents a company that developed independently of Standard Oil during the early twentieth century, it was not alone in Texas. In a way, the Gulf Oil Corporation illustrates a greater success story than Texaco because it operated for several years quite independent of any former Standard Oil Company.

Gulf Oil Corporation

The Gulf Oil Corporation started with the discovery of an oil well in the Spindletop field of coastal Texas in 1901. The founders borrowed the initial capital from the Mellon and Sons banking house in Pittsburgh, Pennsylvania. Later the Mellon brothers would become involved as partners in Gulf. The discovery of Spindletop established Texas as a major oil source, but not as a major market for consumption. The development of an oil industry in Texas posed a challenge for wildcatters because Texas had a sparse population and local demand for petroleum products was limited. Kerosene proved to be largely the only demand, which did not utilize even a fraction of the oil being pumped from Spindletop.[9] The oil field appeared to be an ever-flowing source of crude oil until 1904 when production levels dropped from about 17 million barrels a day to 10,000 barrels a day. Fortunately, oil had been found by independent oilmen in Oklahoma, which presented itself at the perfect time.

In 1907 the Gulf Oil Corporation officially came into being when the Gulf Pipe Line Company, Gulf Refining Company of Texas, and the J. M. Guffey Petroleum Company merged. During the 1910s and 1920s, Gulf Oil expanded its subsidiaries and, thus, its operations overseas to Mexico, Venezuela, and Kuwait. Also, technological advancements revived Spindletop allowing it to reach production levels on 10 million barrels per day well into the 1980s. After weathering the Great Depression in the 1930s, it began to expand its operations into countries considered high-risk, such as Angola and Cuba.[10] Several years later, in 1984, Gulf merged with Chevron Corporation to become a subsidiary.

ConocoPhillips

With its headquarters in Houston, Texas, ConocoPhillips engages in the exploration and production of oil and natural gas as well as commercial activities in over forty countries. The company formed in 2002 through the merger of Conoco Inc. and Phillips Petroleum Company. Conoco formed in 1875 as the Continental Oil and Transportation Company. This Utah based company gained its wealth through the distribution of coal and kerosene throughout the western United States. In 1885 Standard Oil bought out the company. It remained under the Rockefeller umbrella until 1911. In 1928 Continental Oil took over Marland Oil, a Pennsylvania company, which marked the formation of Conoco, mainly a refining and petroleum retail company. After World War II, Conoco expanded its interests to move overseas and establish service stations throughout Europe.

Meanwhile in Oklahoma, Phillips Petroleum Company began in 1905 when Frank Phillips and his brother struck oil. The company officially came into being in 1917 and began marketing gasoline in service stations in 1927. In 1952 Phillips received approval from the U.S. government to drill in Alaska. Throughout the 1940s and 1950s Phillips developed its offshore drilling operations and, in 1955, set the world record for the oil well drilled farthest offshore, 40 miles. During the late 1960s to 2002, Phillips engaged in oil and gas production in the North Sea, the Timor Sea, and China. Since the merger of ConocoPhillips, the company has established itself as one of the world's most profitable companies.

Ente Nazionale Idrocarburi

ENI began as an Italian state-owned oil company in 1953 intending to promote its national interests within the international oil industry. It acted as a holding company. It included the original Italian national oil company, Azienda Generale Italiana Petroli (AGIP), which was created in 1926, as well as refining companies and a natural gas company. Between 1970 and 1975, ENI formed AGIP as its marketing subsidiary and carried out operations within Italy and North Africa. ENI stood out from all other oil companies because it encouraged nationalism among African countries and offered cutting-edge profit sharing (75 percent to 25 percent) that favored the host countries. The company saw itself as a partner, not an exploiter. It often bought and sold low-priced oil from the Soviet Union, showing that it did not necessarily align itself with the United States and Europe during the cold war. Not surprisingly, the major oil companies viewed ENI as playing the role of the underdog within the international oil industry. ENI refused to comply and participate with the major oil companies in their efforts to exploit and extract oil from Italy's own reserves and from those of unstable, colonial African countries.[11] Some argue that this philosophy has allowed the company to expand and flourish. In 1992 ENI transformed itself into a joint-stock company. In 1997 AGIP and ENI began merging; with this reorganization came Snamprogetti in 2000 and Agip-Petroli in 2003. Today ENI operates in over sixty countries throughout the world, making it one of the most successful and unique oil companies.

TotalFinaElf

TotalFinaElf is the product of several decades of mergers. As the name suggests, it is comprised of three major components that came together in 2000 into one large company. The largest component of the company, Compagnie Française des Pétroles (CFP), formed in France in 1924. Four years after its formation, the French government became a major owner of the company, holding around 30 percent, in order to pursue its own political interests and secure overseas supplies. Its operations took place primarily in the Middle East, more specifically Iraq. For many years CFP represented the largest national oil company in Europe. The company found the first oil field in Iraq in 1927 as part of the Iraq Petroleum Company. In 1954, CFP formed its marketing subsidiary company, Total. Six years later, CFP absorbed another French oil-marketing company, Omnium Français des Pétroles. In the 1980s, CFP expanded and changed its name to TotalCFP, and then in 1991 simply to Total. Up to the mid-1990s, the French government continued to hold a share in the company, until it was reduced to less than 10 percent. Free to find new partners, Total brought in the Belgian company PetroFina and became TotalFina in 1999. One year later, TotalFina merged with a French company, Elf Aquitaine S.A., to form the TotalFinaElf we know today. TotalFinaElf operates in more than 130 countries throughout the world.

Lukoil

Lukoil is Russia's largest oil company and the twelfth largest oil company in the world. Lukoil formed out of Russia's privatization policy, which divested the country of its public assets. A presidential decree in 1995 formed several oil companies, of which Lukoil is one. Each company that the Russian government created had a specific role to fill: Yuko's primary activities included oil and gas exploration, production, refining, and marketing. Since its formation, Lukoil has worked closing with ConocoPhillips in the far north. Overall, Lukoil produces almost 20 percent of Russia's crude oil. Lukoil's fields are primarily located in western Siberia in the Timano-Pechora and Yamal regions, which total almost 16 billion in crude oil reserves and more than 24 Tcf in natural gas. Lukoil also takes part in production projects in overseas locations within eastern Europe, Africa, and even South America. In addition to its four refineries within Russia, it also operates refineries in Romania, Bulgaria, and Ukraine. The company markets its products all over the world and employs more than 100,000 people to run its day-to-day operations. What makes Lukoil remarkable is that it has been in operation for only ten years and already ranks among the largest oil companies in the world. Also, it has played a crucial role in turning Russia into a competitive oil-producing country.

In this brief examination of the formation and current operations of the major and minor oil companies, a clear pattern emerges. In each company, the national government played an active role in providing financial and political assistance. This occurred regardless of time period or origin of the oil company. In the case of Britain's British Petroleum and France's CFP, the government took partial ownership in the company to pursue its own political and economic interests. In some cases the companies began as outright state-owned companies, like Eni, that pursued oil interests dictated by the national government. What the formation of these companies indicates is the undeniable role of the national government. Petroleum represents such an integral part of a nation's well-being that production and trade cannot be left to risk-seeking independent companies.

For the most part, oil companies operate independently from the national government. They rarely act in opposition to it. In fact, most companies have an amicable relationship with their government. This is because they depend heavily on each other. The government needs a secure supply of oil for its country and the oil company needs protection during its operations. Especially in the years of oil field discoveries, governments often pushed oil companies into a production project as a way of gaining an international political foothold. For example, the United States pushed its oil companies into the Middle East after World War I. By making sacrifices to satisfy the government's demand, the companies were rewarded. During colonialism, the British government allowed only British companies to operate in its colonies. This gave British Petroleum a distinct advantage over its competitors. By the same token, it is believed that the U.S.-led invasions within the past few years into politically hostile countries such as Iraq and Afghanistan paved the way for U.S. oil companies to set up new operations. This corporate-government relationship does not always stay positive, though. During the oil shocks of the 1970s, the U.S. government threatened to nationalize Exxon based on a belief that Exxon, along with other oil companies, caused a severe increase in oil prices (see Chapter 4). It was believed that since the beginning of the global oil industry, oil companies have colluded in manipulating prices and supply, which often harmed governments and consumers alike. At the same time, however, they maintained a public commitment to the principles of competition.

OIL COMPANIES: CARTEL OR COINCIDENCE?

Until the emergence of state-owned oil companies in oil-producing countries as well as the formation of OPEC in 1960, it was widely believed that the major oil companies functioned as a *cartel*, an organization of firms with the purpose of attaining monopoly power through regulating the members, because only a few companies controlled most of the world's oil resources and operations. Also, the major oil companies came from the

Rockefeller trust, and the public always suspected that they never completely separated into independent entities. Furthermore, the major oil companies controlled not only the production, but also the refining, transporting, and marketing of oil all over the world. Only recently has oil production occurred that does not involve all or most of the major oil companies. As a way of reducing the risks in the high-stakes business of oil, the companies sought to reduce the competition they faced. They did this by forming joint-stock companies and making agreements over territory, prices, and production levels. The result was imperfect competition, in which the major oil companies acted as cartel controlling the world's oil and gas, leaving other companies unable to compete.

Those who see the oil companies as a cartel argue that if competition truly existed among the oil companies then large joint-stock companies such as Aramco in Saudi Arabia would not have formed. It is believed that cartel members commonly use the formation of legitimate businesses under an unrelated company name to create a false sense of competition for the public and to skirt laws. While there are several examples of these joint-stock companies, Aramco is unique in that the company was comprised of most of the major oil companies acting as one company.

The formation of Aramco began in 1930 with prospecting for oil in Saudi Arabia. The king, disenchanted with the former British rulers, but eager to begin oil production, invited in SONJ, Gulf, and SOCAL. Exxon and Gulf declined, not realizing the potential of Saudi Arabia's oil reserves. In 1934 SOCAL accepted and paid 50,000 gold pounds up front. In exchange, SOCAL received exclusive exploration and production rights for sixty-six years. SOCAL then transferred the agreement into its new company, the California Arabian Standard Oil Company (CASOC). Once Saudi Arabia showed promise, SOCAL realized that it did not have the capital or marketing ability to fully utilize the country's resources. In 1936, Texaco, which had developed strong markets, bought half of SOCAL's concessions. In 1946, CASOC invited three other oil companies to expand its operations in Saudi Arabia. Subsequently the company changed its name to the Arabian American Oil Company (Aramco). This joint-stock company was divided into four shares: SOCAL had 30 percent, SONJ had 30 percent, Texaco had 30 percent, and Mobil had 10 percent.[12] Aramco operated under joint control until Saudi Arabia nationalized it in 1979. Of the four companies, three were from the Rockefeller trust. The case of Aramco illustrates the readiness for cooperation among these companies. One of the characteristics of a cartel, though, is collaboration behind closed doors.

Cartels receive a bad reputation because they often make secret agreements, carrying them out without public scrutiny. Consequently, the public feels the impact of their decisions without warning. Two famous agreements, exposed many years after they occurred, reveal that the major oil companies met to make secret agreements. In 1928 SONJ, Royal Dutch/Shell, and Anglo-Persian (BP) began to hold secret conferences. One meeting in particular indicates early efforts to control the industry. The heads of the companies met in Achnacarry Castle in Scotland to hunt and negotiate. They met to discuss the problem of overproduction, which would lead to a decline in prices. From the conference, the oilmen drafted a statement called the Achnacarry Agreement. The principles of the agreement included cutting production levels to a negotiated level, sharing facilities when possible, acquiring new facilities only when absolutely necessary, and refining from excess oil production. In 1930, the companies reexamined the failing agreement and implemented new guidelines in the Memorandum for European Marketing. This agreement called for the establishment of production schedules and the creation of sales quotas. The three companies also agreed on a price level for oil sales in Asia. As quickly as they agreed, each company established side agreements, thus causing the Achnacarry Agreement to collapse. As a result, the agreement never reached full fruition and ended in roughly 1939.[13]

As much as the oil companies attempted to work together as a cartel, however, they followed the basic pitfalls of being one. Scholars view cartels as fundamentally flawed. Each member will disregard the agreed rules, given time, because they are not legally enforceable, especially when the agreement involves cutting production or creating geographic limits. It goes against the fundamental nature of the business world, which is based on success through competition, not cooperation. There are numerous examples of companies intentionally undercutting one another. Nowhere is this more evident than when oil has been discovered in a previously unknown oil field. For example, after the fall of the Ottoman Empire in the Middle East, oilmen sat down and agreed on how the region should be shared among them, using a red pen to indicate the divisions. The members of the Turkish Petroleum Company (TPC: Royal Dutch/Shell, British Petroleum, and CFP) bound themselves not to operate, except through the company, within the area marked on the map by the red line. This area included almost all of the former Ottoman Empire (except Egypt and Kuwait). In areas within the red line, companies from the United States could bid on subleasing a territory, but essentially had to seek permission or include TPC in their activities. In July 1928, the Red Line Agreement was formally signed. This agreement granted, unknowingly at the time, the largest oil-producing region (primarily Saudi

Map 2.1
Map of Red Line Agreement

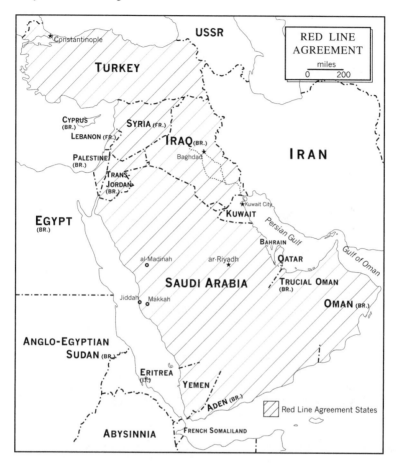

Arabia and Iraq) to non-U.S. companies. Only after the U.S. government intervened did the other companies allow Exxon into their plans.

Whether oil companies operated, or continue to operate, as members
of a cartel, these companies achieved success largely because of their ability to balance competition and cooperation among one another. Furthermore, they have perfected the art of maintaining independence while
keeping their governments as close allies. The history of these oil companies appears contradictory in that the same competition that drove each
company to expand aggressively into the far corners of the world forced
them into business agreements and alliances with each other. In this re-

gard the differences between the major and minor oil companies fade. The major oil companies may run their operations all over the world, but the minor oil companies play a significant role within the international oil industry. Today the distinctions between the major and minor companies have largely blurred as they operate under virtually the same principles of competition and a quest for profit maximization.

FURTHER READING

Only a few works exist on oil companies that go beyond pictorial histories—even fewer that were written within the past twenty years. For a good overview of the oil companies up to the 1970s, see Anthony Sampson's *The Seven Sisters: The Great Oil Companies and the World They Made* (New York: Viking Press, 1975), and "The International Petroleum Companies" in Edith Penrose's *The Large International Firm in Developing Countries: The International Petroleum Industry* (London: George Allen and Unwin, 1968). For the history of Royal Dutch/Shell from the company's perspective, see Stephen Howarth's *A Century of Oil: The "Shell" Transport and Trading Company, 1897–1997* (London: Weidenfeld and Nicolson, 1997) and F. C. Gerretson's *The History of Royal Dutch* (The Hague: Royal Dutch Petroleum Company, 1955), which covers the company's operations until 1914 in three volumes. For the history of British Petroleum, see Ronald W. Ferrier's two-volume *History of the British Petroleum Company* (New York: Cambridge University Press, 1982), which covers the company's history until the 1940s. James Bamberg's *British Petroleum and Global Oil, 1950–1975: The Challenge of Nationalism* (Cambridge: Cambridge University Press, 2000) covers British Petroleum's years in the face of nationalism, particularly in the Middle East. For a detailed history of Standard Oil up to the mid-1950s, see Ralph W. Hidy and Muriel E. Hidy's, *History of Standard Oil Company (New Jersey)*, Vol. 1; *Pioneering in Big Business, 1882–1911* (New York: Harper and Brothers, 1955); George S. Gibb and Evelyn H. Knowlton, *History of Standard Oil Company (New Jersey)*, Vol. 2; *The Resurgent Years, 1911–1927* (New York: Harper and Brothers, 1956); Henrietta M. Lawson, Evelyn H. Knowlton, and Charles S. Popple, *History of Standard Oil Company (New Jersey)*, Vol. 3; *New Horizons, 1927–1950* (New York: Harper and Row, 1971); and, Bennett H. Wall, *A History of Standard Oil Company (New Jersey)*, Vol. 4; *Growth in a Changing Environment* (New York: McGraw Hill, 1988). On the formation and activities of Eni up to 1963, see Charles R. Dechert's *Ente Nazionale Idrocarburi* (Leiden: E. J. Brill, 1963).

Chapter 3

The Oil-Producing Nations

The potential wealth of the Nation, wretchedly underpaid native labor, exemptions from taxation, economic privileges and Government tolerance: these are the factors that built up the boom in the Mexican petroleum industry.

President Lázaro Cárdenas, March 18, 1938[1]

The relationship between an oil-rich country and its oil industry is complex. International prestige and oil revenues contribute to the economic and political elevation of a country. On the other hand, they have brought a rash of problems into often already unstable countries. In many cases, oil discoveries and developments occurred amid political instability and territorial uncertainty. With the exception of a few cases such as Norway, the major oil-producing countries are considered developing countries, which means that they are in the process of working out the social conflict, economic instability, and political turmoil that stand in the way of their joining the ranks of the developed countries such as the United States and those of Europe and the British dominions. It is widely believed that with the right combination of national commitment, leadership, and government policy, a developing country can improve its situation and standing. What characterizes developing countries is their history of foreign occupation, low economic growth, and deficient government leadership—often in the form of military dictatorship. Developing countries also often share the common problem of ethnic conflict and poor social services. In many cases, civil societies in these countries suffer from poverty and government repression. In developing countries, their situation has only become exacerbated by the presence of oil.

There are three phases through which an oil-producing country typically passed as it gained control of its oil industry. The first phase was the arrival of the foreign oil companies. In most cases, the developing country was aware of oil seepages in the ground but was unaware of the amount of oil available or was unable to extract it from the ground. Unaware of the future implications, the developing country accepted the oil company's offers, with few or no rules in place regarding land and subsoil ownership, royalties, and taxation, which ensured a large stake in the industry. Depending on the country, this phase took place between the 1880s and 1950s. The second phase began when the oil-rich country realized that large profits were coming from its oil industry and, more importantly, were leaving the country in the hands of the oil companies. At this point emerged the country's rising desire for control over its oil industry. This typically took place in the 1960s and 1970s, with the exception of Mexico, where it took place in the 1930s. The third phase is the dilemma in which a country found itself when it realized that total control of its industry was not possible in the immediate future. This is where many oil-rich developing countries find themselves today.

It is not oil per se that causes problems for an oil-rich developing country, but what the endowment brings. With the exception of Norway, the discovery and development of oil in a developing country occurred in much the same way whether it took place in the 1910s or the 1950s. A primarily agricultural country with a weak political system in place is approached by a multinational oil company, which wishes to explore and produce whatever oil it finds. The company and the host country form production agreements, which allow both parties—in theory—to benefit.

PRODUCTION AGREEMENTS

A production agreement is a contract between an oil-rich country and a foreign oil company. This agreement grants to a foreign oil company an area of land or water to develop. Over the years, production agreements have changed dramatically. Until about the 1960s, an oil company operated in a country with little direct intervention by the host government. A concession was granted to a foreign oil company, which allowed it to develop new oil fields. Through a *concession*, oil companies often received access to large tracts of land for long periods of time, and received exemption from taxation. For example, in Venezuela during the first half of the twentieth century, concessions were granted to the major oil companies for up to ninety-nine years.[2] In exchange for this, the oil companies paid a percentage of their profits, known as a *royalty*, to the host government. The host country received a set percentage of the oil company's profits (usually 12 to 20 percent), which was based on the *posted price*, or the price predicted by the oil companies. Once the agreement was signed, the host country would find that the *market price*—what the barrel of oil actually was sold for on the international market—was higher and would have brought them greater royalties. This became a major source of tension between the oil companies and the host countries, because the latter felt swindled.

Since the 1960s, the old form of concession has been replaced by a more modern concession along with other forms of production agreements. Tracts of land are generally divided into numbered *blocks* that are distinguished by lines of longitude and latitude. Blocks are most commonly used for assigning offshore oil and gas fields. The period of time granted in a concession agreement has been significantly reduced. For example, royalties were raised and concession periods were shortened considerably in Venezuela starting in the late 1950s. The Venezuelan government granted concessions that allowed for six years' exploration and thirty years' exploitation. Any land unused during the exploration period was returned to the government.[3] The concession rules for Venezuela were fairly standard for concessions of that time. Today, there are three basic types of production agreements, which provide the company with the right to explore, produce, transport, and sell petroleum within a specified period of time and within a specified region.

The most important aspect of production agreements for the host country is the right to collect taxes and royalties. The first type, *the modern concession agreement*, offers the foreign oil company the exclusive rights to every aspect of its oil industry for a period of time (generally twenty to forty years). State officials, under this agreement, are primarily engaged in the enforcement of laws protecting the environment and workers. The concession agreement allows for a *relinquishment clause*, which forces a foreign company to return developed land back to the government. For example, in Norway a concession last six years and includes the possibility of renewal. A relinquishment clause requires that 50 percent of the area granted be returned after the concession expires.[4] As long as the companies operate within the nation's guidelines, the state does not intervene. The *participation sharing agreement* involves the creation of joint ventures between the oil company and the country. It can take several different forms, which are defined in the contract. The final type, the *cont-*

ractual agreement or *service agreement*, is primarily used by countries that are reluctant to allow a foreign company significant involvement in their oil industries. Through this agreement a company agrees to perform specified services for a flat fee for a state-owned oil company. These services often include operations requiring highly skilled labor and technical assistance to conduct geophysical studies or production in hard-to-develop areas. While production agreements are voluntarily made between the oil companies and the host governments, mutual satisfaction is not always ensured. More often than not, resentment and dissatisfaction emerge on the side of the host country.

THE DUAL BUILDUP

Essentially, two situations occur simultaneously: the buildup of a country's oil industry and the country's buildup of frustration toward the oil companies. The oil-producing country's frustration toward the oil companies comes from a conflict in expectations and interests. The expectation of an oil company operating in an oil-rich country is simple. Its primary concern, as with any business, is the maximization of profits and the minimization of risks involved in the process. Beyond the usual risks of developing oil fields in challenging terrains and climates, oil companies face political risks. An oil company expects to conduct its business with little interference from the local community or government. As long as there is oil to produce and sell, the company will remain, but when a field dries up the company does not feel compelled to stay in the host country. Until recently, the company did not view it as its duty to invest in a country beyond its own needs. The host country, however, does not hold the same expectations. The host country expects that, upon discovery, its oil will bring inexpensive oil and employment opportunities to its people, and more revenue to the government. Also, it expects that the company will respect the environment, the people, and the government. More important, however, the host government expects that oil companies will honor profit-sharing agreements and accurately report prices, production levels, and profits. Host countries also have expectations that they recognize only *after* the oil companies arrive.

Oil-rich countries did not always welcome the social and political changes that came with the oil companies. From the public's point of view, a highly lucrative industry arrived in their country, was dominated by foreigners and, for the most part, offered nothing positive to the local community. Host countries complained about numerous social changes that included the disruption of traditional values and ways of living. In many cases, local communities saw the construction of nice camps, roads, and never-before-seen modern technology, coupled with excessive alcohol consumption on the camps, destruction of the local environment, and the accumulation of large amounts of human and industrial waste. Oil fields do not always exist in uninhabited places. Communities were forcibly removed and people were dislocated. Arable land was taken for oil production. The arrival of an oil industry often created towns or urbanized already existing ones. Furthermore, oil companies often developed self-sufficient company towns in which the foreign oil workers lived "physically, economically, and socially apart" from the local people. In the town of Tampico, which became the major oil-refining center in Mexico, foreigners lived in elegant homes on top of a hill where they enjoyed a view of sea. The Mexican migrant workers, however, lived in crowded makeshift homes along the banks of a lagoon on the outskirts of town.[5]

Many dislocated people from oil-rich rural areas flooded urban centers in search of a new home only to find crowded, unsanitary living conditions, and no jobs. This situation increased acts of crime and vice. The oil industry did not appear to bring employment opportunities for many people, but when it did, those opportunities came with a social cost. To work for the oil companies, people often had to learn English, study abroad to receive the training they needed, and adopt the company's lifestyle. In addition to changes derived from employment, local people also complained that the oil companies encouraged racial discrimination.

The public also complained that oil companies fostered political instability. Two opposing trends occurred throughout oil-rich countries: collusion and resentment. Evidence existed that the elite communities and the foreign corporations colluded to exploit oil to their advantage without regard for the nation as a whole. Dictators used the wealth and power of the oil companies to maintain their positions. As a consequence, public grievances toward the oil companies and their rulers became intertwined. At the same time, people who did not benefit from the oil companies viewed them with extreme distrust and contempt. By many, the foreign oil companies were perceived as "heartless new conquerors" and were often depicted in political cartoons as an octopus with each tentacle representing an oil company clutching its prey.

These images and ideas became deeply embedded in the social consciences of host countries. It has been argued that any foreign company, good or bad, easily becomes the target of national frustration. Oil companies operating in a foreign country are often blamed for economic downturns and political conflict, which may or may not involve them. This is because at the core of every nation lies its desire to stand on its own economically without relying on foreign investment, loans, or technical expertise. In Islamic countries, this desire takes on a strong religious component. Islam includes a fundamental principle that calls for self-sufficiency in every aspect of life. To depend heavily on outsiders represents a failure in following Islam. Consequently, both Muslim and non-Muslim countries sought to take control of their oil industries.

One of the primary motivations for actively confronting the oil companies was profit sharing. The oil companies primarily structured a country's oil industry to export, which meant that the oil companies' operations depended on market forces. The oil companies, without consulting the host country, cut production or increased production as they saw fit. As mentioned earlier, the host country received a percentage of the oil companies' profits based on the posted price. This meant that even as the market fluctuated and the price of oil rose and fell, the oil companies always paid the same amount of royalty to the host country. The host country often found that the market price was higher than the posted price, and, consequently, lost potential revenue. This became a serious problem as new fields and markets were discovered throughout the world. As production increased, host countries began to press for greater returns.

Before moving toward outright nationalization, oil-rich countries adopted various forms of legislation designed to allow the host country to become more involved in the industry's operations as well as to keep more of the revenue earned from oil within the country. They either attempted to change the terms of existing contracts or vowed to create more favorable conditions for future contracts. Governments asked for higher royalties or taxes, as we shall see in the cases of Venezuela and Saudi Arabia in the 1950s. They also offered shorter exploration contracts and smaller tracts of land. For example, during the early years of oil exploration in Venezuela, the government in 1907 granted concessions of at least 125,000 acres. By 1943, concessions had shrunk to 25,000 acres. If an oil company failed to find oil within a specified amount of time, the land was returned to the government. Also, oil-rich countries began to set aside land for their own exploration as national reserves, which were not open for concessions. The most difficult and frustrating factor for a host country to understand and control was how the oil companies calculated the production levels that determined total costs, taxes, and prices. Oil companies, naturally, kept their records closed to the public and developed highly complex and intricate ways of determining when and how calculations were made. For example, Saudi Arabia saw that Aramco measured the volume of crude oil at the wellhead so that spills would not be counted as oil produced and, thus, factored into taxes paid to the host government. Once this practice was discovered, Saudi Arabia passed legislation to ensure that all oil, spilled or not, was factored into the calculated production levels. Despite these efforts, the oil-producing countries always mistrusted the oil companies and never felt completely satisfied with their level of involvement in *their* industry.

TAKING CONTROL: EXPROPRIATION AND NATIONALIZATION

Nationalization of foreign oil companies' operations primarily occurred in the 1970s, with the exception of Mexico and Russia (Figure 3.1). The method of nationalization differed from country to country. Taking into account their economic, social, and political situations, each country took the approach that best suited it. Mexico, for example, used an immediate and dramatic method, while Saudi Arabia adopted a more gradual, amicable approach that more closely resembled renegotiations. All of these countries *nationalized* the oil industry, which means they took over all or most aspects of the oil industry that foreign companies had previously controlled. Often in the case of nationalization, the foreign equipment, personnel, and operations are replaced by their national equivalents; elements that cannot be replaced are purchased. Nationalization also includes the formation of a national oil company, which actively operates its own oil resources. Mexico's nationalization, however, is more commonly referred to as an *expropriation* because it was an overnight seizure

Figure 3.1
Nationalization of the Oil Industry

Country	Date
Russia	1920
Mexico	1938
Iraq	1972
Saudi Arabia	1976
Venezuela	1976
Nigeria	1978

of foreign properties and a hasty removal of foreign workers without immediate compensation. Mexico's was not the first expropriation, but it was one of the most important. The first outright expropriation of a foreign oil interest was of Standard Oil in Bolivia, nationalized in March 1937. The Mexican nationalization, however, was far more significant because it occurred during a time of major expansion within an oil-producing country. Also, at the time of nationalization Bolivia produced roughly 2,500 bbl per year, whereas Mexico produced five times as much in a single day during the mid-1930s.[6] Mexico proved that nationalization was possible, which caused the idea to resonate through other oil-producing countries from the 1950s to 1970s.

Mexico

The story of Mexico's nationalization began with four prominent men, and their companies, who dominated Mexico's early oil industry. First, in the late 1880s, William H. Waters and Henry Clary Pierce arrived from the United States with their Standard Oil affiliate company, Waters-Pierce Oil Company. This company focused not on production but, instead, on building refineries and supplying oil to Standard Oil. Second, Edward L. Doheny from the United States arrived in 1901 and formed the Mexican Petroleum Company and the Hausteca Petroleum Company. Finally, a British man, Sir Weetman Pearson (who later became Lord Cowdray) arrived through specific invitation by the Mexican president at about the same time as the two oil companies. By 1903 Mexico had become a commercial oil producer. In 1908 Pearson discovered a tract of oil fields in eastern Mexico known as the "Golden Lane" and formed his Compañía Mexicana de Petróleo el Aguila (referred to as El Aguila). Since Pearson had a special relationship with the president, he was able to create a company that included members of the Mexican elite. For this reason, the company presented itself to the public as a national company, with only Mexico's interests in mind. In reality, the national slant became largely a competitive strategy against Waters-Pierce. Competition among the three companies persisted until the Mexican Revolution, when major oil companies arrived.

The period during the Mexican Revolution represented the Golden Age of oil in Mexico; oil production doubled and more and more oil companies arrived every year. Among many small companies came Royal Dutch/Shell, which purchased Pearson's El Aguila in 1919, and SONJ, which bought the independent British Compañía Petrolera de Transcontinental in 1917. Gulf Oil and Texaco also arrived, but Mexico's industry quickly became dominated by Royal Dutch/Shell and SONJ. By 1920, foreigners owned or controlled virtually every acre of potentially productive oil in Mexico. The country also quickly became the supplier for 25 percent of the world's oil.

At the time of discovery, the political situation in Mexico was at a breaking point. Porfirio Díaz earned a reputation as a harsh dictator and opposition to him emerged. For the foreign oil companies, however, he served as an ally who encouraged their investment and supported their operations. The extraordinary concentration of Mexico's wealth in the hands of foreigners no doubt fueled public dissatisfaction. Popular discontent prompted the Mexican Revolution, which lasted from 1910 to 1920. Although it began as a political movement to overthrow the dictator, it became a violent expression of grievances. At the same time, Mexico's oil industry expanded and frustration toward the major oil companies escalated. By 1920 almost all of Mexico's oil was being exported and was in the hands of foreign oil companies. Mexico's government became extremely frustrated with the issue of property laws and taxation. The Mexican government attempted to implement tax increases in 1917 and 1921, but the oil companies protested. Also, as part of Mexico's new postrevolution constitution, all petroleum was publicly owned and oil companies were required to register their operations, thus turning their holdings into concessions within three months. Thus, the oil companies no longer owned the land and only had production rights to it for a specified amount of time. Mexican landlords and the oil companies refused. The Mexican government then passed the 1925 Petroleum Law, which required the oil companies to register within twelve months and granted concessions without time limits. The oil companies agreed, but tensions developed between the companies and the government over two other issues—labor and oil conservation.

Mexico exported a great deal of oil, but domestic consumption increased rapidly, causing concern among its leaders. In order to ensure domestic supplies, the Mexican government formed Petromex in 1933. Petromex was intended to be a joint venture between the government and private Mexican companies, but the government eventually owned the whole company. The goal of the company included developing new oil-rich territories owned by the government, reducing exports, and supplying the country with oil at subsidized prices. In 1936, Petromex became a totally state-owned company that produced, refined, and distributed petroleum products within the country. Shortly after the formation of Petromex, its true purpose as a national company developed.

The oil companies employed thousands of workers to clear land and do the necessary manual labor for producing new fields. By the mid-1930s, Mexican oil workers totaled about 10,000. These workers organized themselves in labor unions under the umbrella of the federal Confederación de Trabajadores Mexicanos, which formed in 1936. In 1937 the oil workers demanded an increase in their wages and better living conditions and went on strike. The workers brought the government in as a third party to ask the oil companies to meet their demands. Mexico's leaders did not feel it was their place to get involved but did not want to risk losing the workers' political support. When the companies refused to comply with the workers' demands, President Lázaro Cárdenas retaliated by passing the Law of Expropriation, which nationalized the seventeen foreign oil companies on March 18, 1938.[7] The government argued that the interests of the companies stood in opposition to Mexico's interests. Virtually overnight, foreign oil workers left, but without any of the companies' records, which had been seized by the Mexican government. The Mexican government paid the companies $130 million to compensate them for their loss. The decision marked a new day of independence. Mexico's expropriation of foreign oil operations shocked the world, particularly the oil companies. Until that day, no one in the world believed such an action was possible. The decision, however, was the easy part. Mexico moved quickly to pick up where the foreign companies left off without causing any serious stoppage in the flow of Mexico's oil. Three months later, the Mexican government formed its national company, Petróleos Mexicanos (PEMEX), to handle the country's oil industry. The expropriation and the formation of PEMEX were not well received by the foreign oil companies.

The major oil companies and their home countries firmly believed that Mexico's industry, as well as its entire economy, would completely collapse within a month. None of these dire predictions, however, came to pass. In the end it did not matter for the foreign oil companies anyway, because the promise of Venezuela's large oil reserves compensated them for the loss of Mexican oil. Regardless, the expropriation and the formation of PEMEX consolidated national pride in a remarkable way. Mexico still celebrates Expropriation Day, and PEMEX continues to be Mexico's sole oil company, controlling all of its industry. After the expropriation, the oil companies rushed to Venezuela, the next largest Latin American oil-producing country. The oil companies did not view nationalization as an international threat; instead, they saw it as an isolated incident with Mexico. What they did not anticipate was the long years of negotiations between them and other oil-producing countries such as Venezuela and Saudi Arabia.

TAKING CONTROL: RENEGOTIATION

Not all countries felt they could risk total and immediate nationalization as Mexico did. Countries such as Saudi Arabia and Venezuela opted to work with the oil companies to negotiate a better financial relationship. Venezuela took a calculated approach by continuing to negotiate with the foreign oil companies until nationalization seemed a viable option. It spent several years researching and debating the idea of forming a state-owned oil company in order to control its industry. When Venezuela decided to nationalize, it did so gradually.

Venezuela

Commercial production of oil in Venezuela began in the 1920s. Since discovery, Venezuela's oil production continued to expand, and by the mid-1940s Venezuela was the world's largest producer outside the United States. Between 1926 and 1947 Venezuela exported more oil than did all of the Middle East. During World War II, Venezuela acted as the main exporter of oil in the world. When major oil companies such as Exxon, Gulf, and Royal Dutch/Shell first arrived in Venezuela, they received a warm welcome from Venezuela's dictator, General Juan Vicente Gómez. Almost all of Venezuela's oil industry was in the hands of the foreign oil companies until the 1960s. SONJ's subsidiary in Venezuela, the Creole Petroleum Corporation, held the largest concession areas, with Royal Dutch/Shell holding the second largest. Between 1936 and 1940 two changes in leadership moved Venezuela closer to a new nationalistic ideology. The oil companies came under attack for their large profits and poorly maintained oil camps, just as they had in Mexico. From 1941 to 1945, Venezuela's president Isaías Medina Angarita demanded higher royalties and taxes from the foreign oil companies. Forced by the wartime demand for oil, the companies agreed. But this did not represent the end of Venezuela's rocky relationship with the oil companies.

Venezuela's government took a nationalist turn with the formation of a radical political party that in 1945 overthrew President Medina Angarita. When the Acción Democrática (AD) came to power, it unleashed its greatest weapon against the oil companies—an economist named Juan Pablo Pérez Alfonzo, who understood the oil industry very well. As the minister of development (and later minister of mines and hydrocarbons), he pressed for a 50-50 share in oil profits. He hoped to increase Venezuela's control over its oil industry as well as expand his country's role in the world oil market. In 1948, Venezuela passed a law making the government a partner in all of the major oil companies' operations. The oil companies agreed because they recognized Venezuela's heavy dependence on oil; thereafter, the oil companies did not view nationalization as an immediate threat. This arrangement, better known as the Fifty-Fifty Agreement, resonated throughout the oil-producing world. In 1949 Venezuela formally approached oil-producing countries in the Middle East to adopt the agreement, and to create a united front among oil-producing nations. More important for Venezuela, it was an attempt to prevent oil companies from moving their interests from Venezuela to the Middle East.[8]

In 1960 the Venezuelan government created the Corporación Venezolana del Petróleo (CVP), the Venezuelan Oil Corporation, a fully owned state oil company. Much to the government's disappointment, the CVP never became the national company it had envisioned. It suffered from overpoliticization and its operations remained small. The same year that the CVP formed, OPEC did as well (for more on OPEC, see Chapter 4). As a founding member of OPEC, Venezuela increasingly came into opposition with the oil companies and the governments that supported them. Also, with the strength of OPEC increasing every year, Venezuela depended on the major companies less. The companies saw that nationalization was imminent and simply hoped for renegotiation of contracts. In 1974 Venezuela took official steps toward nationalization by forming the Nationalization Commission to conduct research and formulate a plan.

After drafting reports on the specifics of expropriation, Venezuela moved forward. In 1975, Venezuela nationalized its oil and natural gas industry and put it under the control of PDVSA, which absorbed CVP. AD organized PDVSA to act as a state company to oversee the fourteen major oil companies that had been nationalized. The companies received $475 million as compensation. Over the next ten years, the fourteen companies were consolidated. Today PDVSA stands among the top national oil companies in the world. Its operations extend outside Venezuela and include offshore operations in partnerships with foreign oil companies.

Saudi Arabia

Exploration for commercial amounts of oil in Saudi Arabia began in the twentieth century. In 1933, the king of Saudi Arabia granted the country's first concession to an engineer from New Zealand, Major Frank Holmes. Holmes worked closely with SOCAL in negotiations with Saudi

Arabia until the company expanded and brought in partners. Shortly after Holmes received the concession, the California Arabian Standard Oil Company (CASOC) was formed and took over the concession. In 1936 Texaco joined the company. The story of CASOC between its formation and 1940 is peculiar because although it existed in legal terms, it was SOCAL that developed Saudi Arabia's oil fields. In any case, five years after signing the contract, SOCAL found commercial quantities of oil. The first oil field they found, named Dhahran, near the Persian Gulf, was a textbook example of a geological dome that contained oil. In May 1939 the first shipment of oil left Saudi Arabia. Exploration since the 1930s spread throughout Saudi Arabia with SONJ and Mobil taking part. In 1944 SONJ and Mobil joined with CASOC, formed the Arabian American Oil Company (Aramco), which operated Saudi Arabia's oil industry for decades.

As production increased in Saudi Arabia and Aramco's profits steadily climbed, Saudi Arabia decided to renegotiate its role in the industry. What prompted the Saudis to action was a report that in 1949 Aramco paid more taxes to the United States than royalties to Saudi Arabia. Saudi Arabia reportedly received only 21 cents per barrel of oil produced, when the barrel actually sold for more than $2. In 1950, Saudi Arabia made its first move toward a favorable system of profit sharing with Aramco. Although Aramco resisted, it eventually agreed to a 20-80 percent arrangement. The following year, Saudi Arabia demanded a fifty-fifty agreement.[9] This arrangement, in theory, stayed in place until 1976. Although the agreement was made, Saudi Arabia's oil minister accused Aramco in the early 1960s of receiving discounts on the oil they sold, making the agreement more like 32-68 percent. The oil companies denied their actions, which inched Saudi Arabia toward nationalization.

While this trend toward nationalization occurred roughly at the same time as it did in Venezuela, Saudi Arabia moved more slowly, careful to not alienate either the U.S. oil companies or the U.S. government. In the late 1970s, the king formed Saudi Aramco to gradually take over foreign-controlled Aramco. Instead of outright expropriation, the king arranged a handover that was completed in 1980. A name change to Saudi Aramco in 1988 reflected the new ownership. The original Aramco still operates in Saudi Arabia, but only as a provider of equipment and technological assistance. Through this royal decree, the kingdom also took a share in Texaco's operations, forming Saudi Texaco. Saudi Arabia's gradual handover is considered one of the more successful nationalizations because it did not turn the oil companies into enemies.

Outright nationalization, like that in Mexico and Venezuela, was not a viable option for Saudi Arabia for numerous reasons. First, Saudi Arabia had closer political ties to the United States than did its oil-producing allies. Also, Saudi Arabia's oil represented over half of the world's oil. Any lengthy delay or complete stoppage caused by nationalization or by negotiations with Aramco would have had a serious impact on Saudi Arabia's economy and the world's oil supply. Also, unlike Venezuela, Saudi Arabia did not have many of its own people trained and working in the business. Plus, Saudi Arabia needed to sell more oil than there were markets for. For this reason, Saudi Arabia opted to maintain favorable relations with the companies that made up Aramco. Saudi Arabia sought to strike a balance between its national interests and its need for the major oil companies.

Thus far we have discussed only countries in which nationalization occurred. Norway represents a unique case in that it had the distinct advantage of watching the wave of nationalization unfold before its oil industry fully developed.

Norway

Norway was able to avoid the pitfalls of poor infrastructure, laws, and regulations. Even before discovery, Norway held an active role in the prospecting for oil. Norway's oil industry began in the late 1950s with the discovery of the Gröningen gas field off the coast of the Netherlands in the North Sea. This discovery led Norway and researchers to conclude that the North Sea represented an untapped oil-rich location. From the beginning, Norway adhered to strict exploration and production laws in order to lessen the impact of oil development on the rest of Norway's economy. In 1963 Norway granted exploration licenses to Phillips and a consortium of Exxon, Royal Dutch/Shell, BP, and CFP. Norway took the development phase in stages by granting exploration licenses first, and then production licenses. In 1969, Phillips Petroleum struck a giant oil and gas field it called Ekofisk. Production from this field began in 1971. Once production began and oil became a major source of revenue, it became apparent that state participation was necessary. In 1972, Norway reorganized its petroleum operations and formed the Den Norske Stats Oljeselskap A/S, or Statoil, to act as Norway's state-owned oil company. Norway did not feel it necessary to nationalize existing operations, but rather to renegotiate contracts, making the state-owned company a partner. Regardless of *how* a national company is formed, the creation of a state-owned company has both advantages and disadvantages.

THE BENEFITS OF A NATIONAL COMPANY

As previously discussed, establishing a state-owned oil company allowed the government total or at least partial control over its industry. The state-owned company, in theory, has the independence to compete with other foreign oil companies; its close relationship to the government helps protect national interests. Oil-rich countries developed and continue to have national companies because of the many benefits they offer the government and the nation's people.

On an emotional level, a country's nationalization of its industry fosters national pride and unity, especially for those who went through decades of colonialism and a fierce battle for independence. This is particularly true where expropriation also occurred. In the case of Mexico, it confirmed the government's support for its laborers. Long-standing social grievances about the negative influences of the oil companies appeared solved. Whether these problems truly disappeared is debatable, but at least the country showed national support for the protection of its people and the preservation of culture. This is particularly true in the case of Islamic countries such as Saudi Arabia and Iraq which sought to preserve their way of life. With national pride came a positive response from government to make national demands a priority.

Economically, domestic demands take precedence over international demands, and petroleum revenues are concentrated into development projects. The national oil company sets out to ensure that fuel supplies reach all parts of the country at prices reasonable to the population. Often the demands of the domestic market are better recognized and satisfied by the national oil company because it is not as preoccupied with exporting its oil as the foreign companies were. For more than fifty years, Pemex satisfied Mexican demand for petroleum products. A national company is free to seek its crude oil from anywhere in the world at the lowest price possible and to avoid the inflated transport prices charged to affiliate refineries by their parent companies. Within the industry, a national oil company has stronger incentives to use national personnel and national goods and services. With this in mind, it can build and operate its refineries, pipeline pumping stations, and distribution terminals anywhere in the country in order to maximize its social returns and stimulate regional development. For example, in 1980 the Nigerian National Petroleum Corporation (NNPC) constructed the Kaduna refinery in the northern part of the country, far away from the oil fields, in order to develop the north and meet Nigeria's local demands. At the same time, the state company can closely monitor and slow the rate at which its reservoirs are depleted. Saudi Aramco keeps a close watch on its reservoirs to ensure they are not overproduced and damaged. As a state-owned company, it often operates under more favorable conditions than a private business because it is subject to lower taxes and has access to government capital on beneficial terms.

While there are obvious advantages attached to having a state-owned oil company, several oil-rich countries within the last ten years have sought to phase them out of their economic strategies. Finding themselves lagging behind in potential production, technology, and available capital, the national companies have slowly opened their industry to foreign oil companies. Within the past few years PEMEX has recently invited foreign companies to develop its offshore oil and gas fields. Although Mexico's constitution banned oil and gas concessions in 1938, Mexico has considered allowing gas concessions to counter an increase in demand.[10] PEMEX is not alone; Venezuela has also taken aggressive moves to open its gas industry. In 2003 the last of six offshore oil blocks located between Venezuela and Trinidad were awarded to foreign oil companies.[11] Why are these two major proponents of state ownership moving toward privatization?

THE DRAWBACKS TO A NATIONAL COMPANY

Although there are large potential benefits to be gained from developing a national company, there are major drawbacks that present roadblocks to its success. After nationalization, the state oil companies often found themselves grappling with three major problems: lack of capital, lack of managerial and technological expertise, and the often restrictive relationship with government.

The largest obstacle to the national company's success lies purely in its fundamental relationship to the government. As mentioned earlier, the national company acts in close connection with the government. The board members of the company and those holding ministerial positions within the government are intimately connected. The spheres of authority and the objectives of the board's members are often not clearly defined. Because national oil companies operate the most valuable form of capital accumulation within the country, control over the oil company by the government is seen as a measure of the level of control the government has over its entire economy. This is why governments keep a very close watch over operations and, as a result, prevent the company from acting as a highly competitive and truly private company. The oil company is often forced to abide by government-set price controls and required to purchase oil from local suppliers, pay high wages, and adopt labor-intensive methods of operation to promote employment. All these requirements force the oil company to reduce its cost efficiency and make it unable to truly compete with independent oil companies. Once the national oil companies cease to make sufficient profits to sustain themselves, they draw from government funds in order to continue operating. The praised national oil company in many cases becomes one of the government's most expensive undertakings. It does not take long before a lack of capital affects all other aspects of the government's oil industry.

First and foremost, the lack of capital quickly catches up with the maintenance of existing oil facilities (such as refineries and pipelines) as well as future plans for expansion. When the country nationalized its oil industry, the state-owned oil company acquired oil facilities in whatever condition they were in at the time of expropriation. In many cases this meant that the company inherited outdated, broken-down, and corroded equipment. More often than not, the state-owned company continued to run the facilities as they were in order to accumulate oil revenue and invest it quickly into national improvement projects. In a push to make its oil more accessible to its people, the country invested in building more refineries and extensive pipeline systems that it later could not afford to maintain. In Nigeria's case, NNPC nationalized the Port Harcourt refinery and used some of its own capital to build a second refinery in Port Harcourt (1985) and one in Warri (1978). By the late 1980s, the refineries showed signs of neglect and began to break down; they have yet to be consistently operable. In 2004 the refineries collectively processed roughly 300,000 bpd. Unable to invest the necessary capital to make them consistently run efficiently and safely, NNPC put its refineries up for sale to private companies beginning in 2002. As of publication, no purchases were made, leaving the refineries to further deteriorate. In 2005 Nigeria began its search for independent companies to construct ten to fifteen small refineries to go on stream in 2008. A similar problem occurred in the employment sector of the oil industry.

From the beginning of a developing country's oil industry—when it was dominated by foreign companies—the host country struggled with the company's hiring practices, which were passed on to the state-owned oil company. The oil companies had a reputation for not hiring citizens of the country in which they operated, instead bringing in their own technicians and engineers to run operations. When they did hire from the local community, it was often to stave off public criticism. They hired local people for often unskilled positions such as security guard. Those that were hired in the downstream sector to clear land and build oil facilities often lived in unsanitary camps separate from their expatriate overseers. While many companies claimed to "indigenize" their operations by training locals for skilled jobs, they did not place local people in influential or decision-making positions.

Oil companies resisted the idea of building a successful industry and then having to transfer the knowledge and skill that made it possible. The oil companies have little incentive to use local manpower, except for the minor jobs of kitchen duty, maintenance, or transportation. Oil companies resisted the transfer of skills, or *transfer of technology*, because the transfer did not further their main objective, which was to extract oil and make profits. When the foreign oil companies left, they took their engineers and technicians with them, leaving the new state oil companies shorthanded and without training programs to generate replacements. After nationalization, state-owned oil companies found themselves not only short of technical expertise, but also short of the capital necessary to create training programs and new positions.

The national companies imported equipment and technological expertise, which negatively affected their budgets. Overall, the transfer of technology moved slowly. More and more students were sent overseas to learn the necessary skills to run the industry. A leading historian on BP stated, "Nationalization may have solved problems of labor exploitation or exclusionary hiring practices, but it created the problem of gathering qualified workers to run the oil industry's day-to-day operations."[12] Most state oil companies had to adopt arrangements that allowed them to nationalize, but also permitted them to take advantage of the technological expertise of a foreign oil company.

In the case of Iraq, the Soviet Union provided the technological assistance that made nationalization possible. In 1961 Iraq passed Public Law 80, whereby Iraq expropriated 95 percent of the foreign-owned Iraqi Petroleum Company's concessions. Having gained public oil fields, Iraq announced in 1964 its decision to form the Iraqi National Oil Company (INOC) to take complete control of its oil industry. Iraq, however, knew that outside help was an absolute necessity. In 1967 Iraq and the Soviet Union signed the Iraq-Soviet Protocol, which committed the Soviet Union to providing technical and financial assistance to INOC. More specifically, the Soviets agreed to help Iraq develop the Rumaila oil field, which is one of Iraq's largest fields. In exchange, Iraq was to pay back the Soviets in crude oil. In 1972, nationalization was complete. Iraqi leaders knew full well that this was only possible because of Soviet assistance.[13] Iraq as well as other oil-rich countries realized the importance of balancing nationalist goals with realistic steps. Creating a national oil company has significant advantages and disadvantages; oil-producing countries had to find solutions that worked best for their situation.

STRIKING A BALANCE

Countries wanted to strike a balance between their interests in economic development and the foreign oil companies' interest in low-risk profits from the first discovery of oil within their country. Many oil-rich countries, however, saw many of their expectations go unfulfilled by foreign oil companies. To improve the situation, the oil-rich countries adopted a viable approach to nationalization that best suited them. After nationalization, the state oil companies found themselves grappling with a lack of capital and technological expertise. For this reason, oil-producing countries sought support and answers from one another.

FURTHER READING

The tension between foreign oil companies and oil-rich governments has generated a substantial collection of books and articles. Works such as Michael Tanzer's *The Political Economy of International Oil and the Undeveloped Countries* (Boston: Beacon Press, 1969) and Edith Penrose's *The Large International Firm in Developing Countries: The International Petroleum Industry* (London: George Allen and Unwin, 1968) look at the relationship between the foreign oil companies and their host countries. For the nationalization of Mexico's oil industry, see Antonio J. Bermúdez's *The Mexican National Petroleum Industry: A Case Study in Nationalization* (San Jose, CA: Stanford University Press, 1963), Roscoe B. Gaither's *Expropriation in Mexico* (New York: William Morrow, 1940), and Laura Randall's *The Political Economy of Mexican Oil* (New York: Praeger, 1989). For the nationalization of Venezuela's industry, see Gustavo Coronel's *The Nationalization of the Venezuelan Oil Industry* (Lexington, MA: Lexington Books, 1983) and Laura Randall's *The Political Economy of Venezuelan Oil* (New York: Praeger, 1987).

Chapter 4

OPEC and International Oil Organizations

Two separate issues were at stake. One had political implications. That was the war and the question of oil as a political instrument. The other was the price of oil. The problem was that these two issues could be easily confused as being the same.

Sheikh Ahmed Zaki Yamani, 1988[1]

At the same that several oil-rich countries nationalized their oil industries and effectively removed the stranglehold of the foreign oil companies, several other events shaped their newfound strength in the global trade of petroleum. The most important event, the formation of OPEC in 1960, created a new forum for tension between the foreign- and state-owned oil companies, with OPEC providing a collective voice for the developing, oil-rich countries. While OPEC grew in strength during the 1960s and 1970s, political tension in the Middle East escalated and supplies of oil in the world became unstable. Shortages and gluts magnified natural price fluctuations of oil products, while the oil-producing countries felt powerless. At the close of the 1970s, several wars took place within the Middle East, an oil price revolution occurred, and the first semisuccessful oil embargo took place. To counter the new power of OPEC and the regional Organization of Arab Petroleum Exporting Countries (OAPEC) that formed in the Middle East, the developed countries (home of the foreign oil companies and most of the world's oil importers) formed the Organization for Economic Cooperation and Development (OECD). The 1960s and 1970s represent two decades of turning points in the history of global oil politics because of the events and changes of power that took place. In this chapter, we will go through these remarkable decades to introduce the major organizations and retell the series of events collectively referred to as the first oil shock, which left no part of the world unaffected.

In the previous chapter, we discussed the buildup of frustrations of the major oil-producing countries with the foreign oil companies. They accused the oil companies of rigging profits over the years by concealing the real figures and facts on production and price. The oil-producing countries wanted to gain control over their own industry, have a say in crude oil prices, and implement policies to rein in profits that went to the foreign oil companies. Although some countries such as Mexico nationalized their industries and formed state-owned oil companies, this process was the exception. Most oil-producing countries by the late 1950s had not formed national oil companies and fully taken control of their industries. Therefore, when a price reduction in 1959 took place, the oil-producing countries began to see a need not only to unify, but also to take decisive action toward controlling their oil industries. The 1960s and 1970s represent the height of frustration among oil-producing countries, which resulted in a dramatic power shift between the foreign company and the host country.

In 1960, the major oil companies became concerned with a glut of oil on the world market, which drove down the price of oil all over the world. They had already cut prices in 1959 and were preparing to do it again a year later. This problem was compounded by the cold war, which had created a bipolar world in which oil companies were forced to pick sides. The Russians had created their own markets in African and Asian countries, where they sold oil at even lower prices than the major U.S. oil companies. While they felt it necessary to lower their prices in order to compete with Russian oil, U.S. companies were locked into the Fifty-Fifty Agreement (see Chapter 3), which guaranteed a 50-50 profit-sharing contract between the oil company and the host government. The agreement also meant that the oil companies had to honor the posted prices on which the agreement had been made. When the price of oil went lower than the posted prices, the oil companies came up short. The largest and most influential company at the time was Exxon. Exxon intended to lower the posted prices everywhere except in Venezuela to protect its already fragile relationship with the country. When the announcement went out, the Middle Eastern oil producers did not respond favorably. Within months the Arab oil-producing states met in Cairo, with observers from Venezuela in attendance as well. They advised the oil companies not to reduce the posted prices without consulting with the governments first. In the meantime, talks between two influential people in the oil business—one from Venezuela and one from Saudi Arabia—resulted in the formation of OPEC.

VENEZUELA AND SAUDI ARABIA

As mentioned in the previous chapter, Venezuela adopted the Fifty-Fifty Agreement in 1948 and appeared to be inching toward nationalization. By 1960 Acción Democrática (AD) had been in power intermittently for almost twenty years. More important, a Venezuelan economist, Juan Pablo Pérez Alfonzo, had been pushing for more drastic measures. Pérez Alfonzo was born in Caracas in 1903 and spent his student years as a political activist. In 1939 he entered the political arena and became the minister of development in 1945. It was through Pérez Alfonzo's work that Venezuela became one of the founding countries of OPEC.

Pérez Alfonzo, as the minister of development (and later minister of mines and hydrocarbons), hoped to increase Venezuela's control over its oil industry and launch his country into the global oil market. The fact that Venezuela had been excluded from the posted price cuts in 1960 made him more determined to unite with the Middle Eastern producers. The position of Venezuela in the late 1950s was unique in that it produced more oil than Kuwait and Saudi Arabia, the Middle East's largest producers, put together, not to mention that Pérez Alfonzo had a reputation as a charismatic, influential person.[2] Pérez Alfonzo had for years believed in *pro-rationing* among oil-producing countries as a way of controlling the markets. The idea of *pro-rationing* was essentially a quota system by which oil producers agreed on how much each country would produce as a way of curbing gluts on the market and, in turn, controlling oil prices. It was not until the 1960s, however, that his radical ideas found support outside Venezuela.

Like Venezuela's Pérez Alfonzo, Saudi Arabia had its own new oil adviser, Abdullah Tariki, with similar radical views. Both men saw the need to unite the producing countries. Tariki was born in 1919 in Saudi Arabia. Tariki studied petroleum geology at the University of Texas and worked briefly with Texaco. He returned to Saudi Arabia in 1948 armed with an arsenal of knowledge on U.S. oil companies. He became the director general of petroleum and mineral affairs in 1957 and was promoted to minister in December 1960. The ideas of Pérez Alfonzo inspired Tariki. In 1962, Sheikh Zaki Yamani took over as oil minister until 1986 and complemented Pérez Alfonzo's vision quite well. He felt that the oil-rich countries should actively join the major oil companies instead of acting in opposition to them. In April 1959, Iran, Kuwait, Saudi Arabia, Egypt, Syria, and Venezuela met and created the Mehdi Pact. Although primarily symbolic, the pact represented a mutual interest in establishing national oil companies, receiving a greater portion of oil revenues, and taking a greater role in determining oil prices. The pact did not receive much public attention and had no legally binding commitments.[3] The failure of the pact, however, was easily overshadowed by the monumental formation of OPEC only six months later.

ORGANIZATION OF THE PETROLEUM EXPORTING COUNTRIES

OPEC was formed in 1960 when five oil-producing nations—Iran, Iraq, Kuwait, Saudi Arabia, and Venezuela—met in Baghdad to examine the control of petroleum by the former Seven Sisters. The founding members sought to implement a collective strategy to gain more control over their industries and to increase their returns from the oil profits made by the foreign oil companies. Today the organization includes almost all of the world's largest oil producers. OPEC is headquartered in Vienna, Austria, which serves as a midpoint among the geographic locations of all OPEC members. The organization holds two conferences a year but often holds additional meetings when necessary. Each member of OPEC is represented by a delegate, usually the country's oil minister. The most publicly visible member of OPEC is the secretary-general, who manages the organization's headquarters in Vienna and organizes OPEC conferences. The current secretary-general is Purnomo Yusgiantoro, who also serves as the minister of energy and mineral resources for Indonesia. The goal of the organization has remained virtually unchanged since its formation (Figure 4.1).

OPEC's main objectives include controlling the price of crude oil and controlling the quantity produced. Its aim is creating a stable market for petroleum producers as well as consumers through establishing price levels and production quotas for each member country. Every year, members meet and agree, in private meetings, on the quotas and prices.

The pricing of crude oil can be quite complicated because not all crude oil holds the same market value, but all crude oil is subject to the fluctuations of the market. In order to develop a simpler system for pricing

Figure 4.1
OPEC Members

Founding Members	
Iran	Saudi Arabia
Iraq	Venezuela
Kuwait	
Other Members	
Algeria (1969)	Libya (1962)
Ecuador (1973–1992)	Nigeria (1971)
Gabon (1975–1994)	Qatar (1961)
Indonesia (1962)	United Arab Emirates (1967)

crude oil, traders established a number of reference points, known as *benchmark* prices, for all crude oils. For international trading of crude oil, Brent from the North Sea is considered the benchmark, while for trade in the United States, West Texas Intermediate (WTI) is used. Other crude oils are priced according to their quality. To add to the already complicated system, OPEC has its own reference based on crude oil produced by its member countries. In general, the price differences between Brent, WTI, and the OPEC basket are small. The *OPEC basket* indicates an average price based on the different types of OPEC crude oils. From this basket, OPEC monitors and maintains the price of oil to stay, until recently, between $22 and $28 per barrel. Price hikes reaching $50 per barrel of crude oil since 2003, however, have pushed OPEC to reevaluate its price range (see Chapter 9). The most important way that OPEC maintains this price range is by setting production quotas so that there is never too much or too little crude oil available on the market, which would adversely affect the market price.

OPEC controls the basket price of oil by adjusting the amount of oil its members pump daily out of the ground. For more than forty years, OPEC has been setting a production quota for each member country within an overall ceiling. The quota for each member country is based on the quantity the country is able to produce as well as its level of recoverable reserves. Saudi Arabia plays a critical in OPEC as the *swing producer*. Because Saudi Arabia is the world's largest oil producer, it can afford to increase or decrease the amount of oil it produces quickly without creating economic problems within the country. Whenever OPEC needs an adjustment in the market, it looks to Saudi Arabia. While this system has been effective for the most part, it is not foolproof. Supposedly, OPEC members agree to meet, and not exceed, their annual quotas. In practice, however, OPEC's members often disregard their quotas. This is where the nature of being a cartel works to OPEC's disadvantage.

One of the unique aspects of OPEC is that it is one of the few international cartels in existence. The entire organization operates on each member's good intentions and promises. In other words, no legally binding agreements are made, and there are no legal repercussions if a member violates an agreement. The formation of world oil cartels is not new. In fact, the actions of the foreign oil companies in the world's most oil-rich countries as a cartel influenced the idea for OPEC. It has been argued that the major oil companies functioned as a cartel because they were the few companies that controlled most of the world's oil resources and operations.

These companies controlled not only the upstream sectors, but also the downstream activities all over the world. They formed joint-stock companies and made agreements over territory, prices, and production levels in order to reduce competition. OPEC functions in much the same way, only more to ensure that its members' interests are met than out of a desire to eliminate competition. As is any cartel, however, it is only as strong as the members allow it to be.

One of the problems with OPEC as a cartel, and of its use of a quota system, is that the cartelized price of oil often exceeds the marginal cost of production for the members of the cartel, which results in a strong incentive to cheat. OPEC members cheat by making secret contracts that go outside the OPEC agreed-upon protocol on quality and price of the oil. Member countries undermine the solidarity of OPEC by regularly producing beyond the agreed-upon targets. Who is actually cheating, however, is generally not revealed to the public. While the consuming countries and the oil companies have waited to see whether the fundamental flaws of forming a cartel would pull OPEC apart, they have so far been disappointed. What has weakened OPEC over the years, in fact, has been internal conflict that, though tied to oil, is not tied to OPEC. Regional tensions among the Middle Eastern countries have weakened OPEC's unity, as has the inability to attract and retain major oil producers from outside the Middle East.

OPEC's weakness has been also been its inability to include three major oil-producing countries, as well as the minor oil-producing countries outside of the Middle East, that would give OPEC a truly international feel.

In the 1990s, OPEC saw the withdrawal of Ecuador in 1992 and Gabon in 1994 because they sought to produce at rates higher than OPEC recommended. For related reasons, OPEC has not been able to bring into the organization Mexico, Russia, and Norway—three of the world's largest oil producers and exporters. Mexico chose not to join because it had already severed its ties to the foreign oil companies in 1938 and did not need to join the "OPEC revolution." Nonetheless, Mexico for the most part has adhered to OPEC pricing policy and production levels since the late 1990s without becoming a member. Norway has acted in much the same manner as Mexico since 1998. Norway did not feel it needed to join OPEC because its industry did not gain serious attention until after the formation of OPEC. Norway was able to learn from the growing pains experienced by OPEC's members. Russia, on the other hand, appears to work with and against OPEC at different times. Since 2000, Russian production has increased significantly. In September 2003, Russia's crude oil output exceeded Saudi Arabia's. This makes Russia a very important producer that often works against OPEC's measures to maintain a fair and stable petroleum market. When OPEC set the price for Russia's oil on the world market in 2003, Russia intentionally undersold. While attempts were made to bring Russia in line with OPEC's interests for a stable market, Russia has expressed no interest in becoming a member.[4] A large part of the reason these countries do not feel it is necessary to join OPEC is that they benefit from OPEC's work whether they are members or not. For those who joined OPEC during its early years, however, becoming a member was being part of the "OPEC revolution."

OPEC encouraged its member countries to make significant structural and political changes to evoke a revolution. For those countries that joined OPEC, their membership empowered them. The organization gave oil-rich countries a platform to discuss their oil industries and share their frustrations with the foreign oil companies. The one wish that OPEC has not been able to fulfill was the desire for Arab oil-producing countries to build a greater alliance among themselves.

ORGANIZATION OF ARAB PETROLEUM EXPORTING COUNTRIES

Several of the Middle Eastern members of OPEC decided to form a separate organization called the Organization of Arab Petroleum Exporting Countries (OAPEC), establishing its headquarters in Kuwait. OAPEC first met in Beirut on January 9, 1968, and consisted of only three members—Saudi Arabia, Kuwait, and Libya. The decision to form OAPEC came in the aftermath of the 1967 Arab-Israeli War.[5] The founding members formed the organization as a way of building regional alliances (therefore adopting an all-Arab membership) and attempting to remove oil as a political weapon and as a factor in global politics. For this reason, the founding members specifically excluded Egypt and Syria, the major actors in the 1967 Arab-Israeli War. Shortly after OAPEC's formation, the political scene in the Middle East changed, prompting the membership to include Egypt, Iraq, Tunisia, Qatar, Bahrain, Algeria, Syria, and the United Arab Emirates. In 1979 Egypt was suspended temporarily, and in 1986 Tunisia permanently withdrew from the organization. OAPEC is run by a council of petroleum ministers, or comparable officials, from each of the member countries. The organization meets at least twice a year or at the request of the secretary-general, who currently is Abdulaziz A. Al-Turki, from Kuwait. OAPEC continues to act as an influential organization today, particularly among Arab oil-producing countries (Figure 4.2).

OAPEC has contributed to the wealth and success of the Middle East's oil industry by creating an Arab tanker fleet, ports, and a service company. The organization has also fostered a solid foundation of cooperation and integration in the petroleum industry of Arab states. While the organization has contributed to the empowerment of the Arab oil-producing countries, it has been accused of undermining OPEC by pulling cooperative interests into an Arab alliance as opposed to a global oil producer's

Figure 4.2
OAPEC Members

Founding Members	
Kuwait	Saudi Arabia
Libya	
Other Members	
Algeria (1970)	Qatar (1970)
Bahrain (1970)	Saudi Arabia (1968)
Egypt (1973)	Ayria (1972)
Iraq (1972)	Tunisia (1982–1986)
Kuwait (1968)	United Arab Emirates (1970)
Libya (1968)	

alliance. What makes OAPEC important to the 1970s is that the membership included those who played a major role in the first oil shock, where oil clearly became the focus of global conflict and was used as political weapon.

THE FIRST OIL SHOCK

During the 1970s and 1980s three major global petroleum-related catastrophes occurred. Collectively they are referred to as *oil shocks* because they were quick, unforeseen occurrences that dramatically shifted oil prices. The effect rippled through the world, leaving no region untouched. The first oil shock took place beginning in 1973 due to three factors. First, the producers, particularly OPEC, were demanding greater participation in the oil industry. Second, there were increasing signs of an oil shortage. Third, the Arab-Israel situation was again heading toward military confrontation. The convergence of the three created an avenue for an oil price revolution and an oil embargo. Even if consumers around the world did not follow politics in the Middle East, they felt the effects of it when petroleum prices soared to unprecedented levels. For example, Saudi Arabian crude oil went from $3 a barrel in 1973 to $36 a barrel in 1980. Two other oil shocks followed in the 1980s, but they will be discussed in greater detail in Chapter 8. The first oil shock represents a period in history when politics and oil came together on a truly international scale. The first oil shock seemed to the average consumer in the United States to be a confusing storm of conflicts in a faraway part of the world involving a complex oil industry that few outside the business really understood.

The first oil shock is best understood as two major events occurring simultaneously. First were the oil price hikes, which took place during the early 1970s, before 1973. These price hikes were the result of a battle between the oil-producing countries and the foreign oil companies. The end result was considered an oil price revolution because OPEC for the first time took control of the price of the oil on the international market. Second, several Arab states used oil as a political weapon in the form of an oil embargo. Frustrated with the United States' support of Israel, they stopped all sales and shipments in 1973. The embargo included only the Arab countries (OAPEC members), which meant that Ecuador, Gabon, Iran, Nigeria, and Venezuela were not involved in the oil embargo. Even Iraq did not favor the embargo as the ideal approach to the problem of Israel. All these countries, however, as members of OPEC, were involved in the oil price revolution. While the two major events are not completely connected, they took place simultaneously, creating the first oil shock. For purposes of clarity, we will discuss them separately.

The Oil Price Revolution

When OPEC formed in 1960, the organization's members maintained the goal of having a say regarding the price of *their* petroleum products on the world market. They were not immediately successful until the 1970s because they needed to gather data, strategize on fixing prices, and determine the best method to restrict production. In the meantime, the oil companies continued to conduct their business on an individual basis with each producer country. They did not perceive the formation of OPEC as an immediate threat. What finally took place in the 1970s was, in fact, a revolution because the change was permanent. Within those two decades, OPEC took over who regulated and determined the price of oil on the international market. In the 1960s, OPEC's members pushed for more control and better prices piecemeal, through individual negotiations with the oil companies operating in their country. By the 1970s, OPEC had not only a clear sense of its mission, but also a plan for how to go about achieving it.

The 1960s represents a push-and-pull decade between OPEC members and the major oil companies. The oil companies focused on the rising global demand for oil and keeping disruptions down, while the oil-producing countries inched their way toward an oil price revolution. Much of the resolutions passed by OPEC during the 1960s called for actions among individual countries. In 1962, OPEC passed three resolutions. The first called on OPEC members to negotiate with the oil companies for the restoration of posted prices to their pre–August 1960 levels. As mentioned earlier, during the late 1950s the oil companies dropped the price of oil repeatedly and unevenly among oil-producing nations. The second resolution called for members to demand that royalties be determined separately from the 50-50 profit-sharing agreements. This way the royalties would be based on total production and not on a portion of it. The third resolution was similar to the second in that it challenged the way the oil companies calculated their payments to the oil-producing countries. Oil companies would be asked to stop deducting expense allowances from the posted prices before determining the taxable income. This would increase the amount of income tax paid to the oil-producing country. These resolutions marked the beginning of many years of negotiations and pressure.

When OPEC members took the resolutions to the respective oil companies, they did not receive a positive response. The companies resisted because the changes would mean a reduction in their profits. After two years of negotiation, the companies agreed on the second resolution that increased profits for the oil producers by $.035 per barrel beginning in 1964. In 1968, the oil companies and OPEC members agreed to phase out the expense allowance completely by 1972 (it was actually eliminated one year earlier than scheduled, through the Tehran Agreement). While these negotiations appeared successful for both sides, a glaring omission existed. OPEC still had not managed to get control of oil prices, which began shifting downward and continued to do so until 1969. The economic situation of the 1970s made addressing the issue of oil prices for OPEC members impossible to ignore.

During the 1970s, global economic crises hit a high note. The price of oil decreased as a result of the United States' decision to discontinue redeeming the dollar for gold as required under the Bretton Woods Agreement of 1944. This led to a depreciation of the U.S. dollar against other currencies. This was a major concern for OPEC because the U.S. dollar was the currency used to denominate oil prices and calculate oil revenue for oil-producing countries. As a result, OPEC went back to the oil companies to adjust oil prices upward as a way of offsetting the loss in the U.S. dollar's purchasing power. Every day the price of oil remained too low represented a major loss for the oil-producing countries in royalties and income tax from the oil companies. Because of the United States' central position in the world economy, its problems impacted countries all over the world. Likewise, events taking place outside the United States only exaggerated the economic problems of slow growth and inflation occurring within the United States.

Three global problems occurred in the early 1970s that adversely affected already unstable oil prices. First, the oil companies underestimated global demand and faced a potential shortage. By the early 1970s, this became a serious concern. If a potential shortage existed where all oil fields, facilities, and transportation routes were running smoothly at full capacity, then anything less than perfect production and transport would create a severe shortage. In 1975, this became a reality when the Trans-Arabian Pipeline, which ran northwest from Saudi Arabia's oilfields to the coast of Lebanon on the Mediterranean Sea, closed. A combination of political turmoil and economic circumstance in Lebanon and Syria as well as conflict in Israel prompted the closer of the pipeline to Lebanon. From the 1980s to the start of the Gulf War in 1990, Saudi Arabia used the pipeline to transport oil to Jordan. Because of Jordan's support for Iraq, however, Saudi Arabia closed the pipeline entirely. The trans-Arabian pipeline to Lebanon represented the oil lifeline for Europe. Replacing the convenience of this pipeline with other forms of transportation drove the price of carrying the oil, and, consequently, the price to the consumer, upward. The oil companies raised the posted price of oil in the Gulf States. Venezuela responded by passing legislation in 1970 that would make prices uniform across oil-producing countries. It also increased the oil companies' tax rate to 60 percent. This marked the beginning of truly unilateral pricing among OPEC members. From this point forward, OPEC acted as an organization with power.

As a collective force, OPEC took control of the industry and oil prices through two major agreements with the oil companies. The meetings that resulted in these agreements were monumental because not only did OPEC members act as one, but the oil companies did as well. All the major oil ministers as well as company heads took part in these meetings. Furthermore, delegates were sent on behalf of the United States and other Western governments to ensure that their needs were met and that the oil companies did not make any rash decisions. In other words, they made sure that the oil companies stayed at the negotiation table. They definitely did not want a shutdown, which was what OPEC promised if an agreement was not made. The two meetings included one in Tehran, Iran, from which emerged the Tehran Agreement, and another in Tripoli, Libya, that concluded with the Tripoli Agreement. The meetings indicated only the starting point of long and intense negotiations between the two groups. In each case OPEC set deadlines, using the threat of nationalization to pressure the oil companies.

OPEC met in December 1970 in Caracas, Venezuela, to solidify and announce its demands to the public. OPEC called for a blanket increase in posted prices and the total removal of expense allowances, which was originally planned for 1972. Also, it proposed that 55 percent the new standard minimum income tax for foreign oil companies in all member countries. Many members, such as Venezuela, went beyond the 55 percent minimum by demanding 60 percent for their country. Previously, it had called for a rate of about 50 percent. One month later, the oil companies responded with a "Message to OPEC," which called for a formal meeting during which all these issues would be ironed out face to face. The ultimate goal of the meeting was to reach an agreement on posted prices and tax rates that would last for five years. At the time, OPEC members represented two general types of personalities, and it was to their advantage to play to these strengths. Therefore, they split into two camps for the purpose of meeting with the oil companies. In Tehran, Saudi Arabia's Sheikh Zaki Yamani acted as the strong voice for OPEC, while in Tripoli, Libya's representative did most of the talking because of Colonel Mu'ammar Gadhafi's unwavering commitment.

The meeting in Tehran took place in January 1971, with a deadline for negotiation set for February 15 of that year. OPEC members from the Gulf States attended, with Yamani as the front man. The negotiations began with OPEC's demand for an increase of $.54 per barrel and the companies' response proposing an increase of $.15 per barrel. On February 14, the oil companies and the Gulf States settled on a five-year agreement giving the Gulf States an increase of $.35 per barrel, with an annual increase of $.05. Furthermore, the prices of heavy crudes (which existed primarily in Venezuela) would increase under a price scale using gravity to determine the value calculated in weight. In exchange, OPEC promised to adhere to the agreement and not initiate an embargo. The same agreement on behalf of OPEC took place one month later in Tripoli.

After the Tehran Agreement, OPEC's more radical and anti-Western members met with the oil companies in Tripoli, Libya, with a negotiation deadline of March 10, 1971. This group of OPEC members represented the countries whose exports went primarily across the Mediterranean (Libya, Algeria, and Iraq). Negotiations went on past the original deadline even though many of the demands by the OPEC members were the same as those resolved in the Tehran Agreement. In April the Tripoli Agreement was signed, whereby the posted price increased and the tax rate increased to 55 percent. In both agreements, the long-standing 50-50 profit-sharing arrangement became 55-45 in favor of the host countries through an increase of the tax rate. Neither of these agreements directly stated that OPEC would control the price of oil; instead, the agreements gave OPEC members greater control over their industry and, hence, increased control of price through tinkering with production levels and tax rates. This placed OPEC in a prime position to handle the rising oil prices that took place in subsequent years.

The 1973 Oil Embargo

Several heated issues preceded the actual embargo in 1973. The primary catalyst, however, stems from a historical tension over the state of Israel in the Middle East. The 1973 oil embargo represented a breaking point in this frustration by the Arab nations over Western support of Israel. Because so much of the oil crisis during the 1970s is wrapped up in a long-standing dispute over the state of Israel in the Middle East, it is important to take a moment to examine it.

The origin of this Arab-Israeli conflict dates back many years. Jews and Muslims in the Middle East have a mutual claim to the region stretching back thousands of years, with deep religious relevance. Three thousand years ago the region was known to the Jews as the "land of Israel" and thus holds many holy sites. This same region is also referred to by Muslims as Palestine because it was settled by the Philistines. It holds Islamic holy sites associated with the Prophet Muhammad. The development of the Israeli state began with the Zionist movement.

Zionism is religious-based movement seeking the return and resettlement of the Jewish people to their historic Palestinian homeland. This movement gained momentum during World War I and gained Western acceptance during World War II. In 1948 the United Nations established the state of Israel and encouraged the diaspora Jews to immigrate. At the same time, 700,000 Arab refugees left the Israeli-occupied areas.

Tension within the region culminated into five wars collectively referred to as the Arab-Israeli Wars, with Arab states fighting over the formation of the state of Israel and the political rights of the Palestinians. In each of the wars, the focal point included the shifting of borders. The West Bank is a region west of the river Jordan and the Dead Sea and includes the historically important East Jerusalem, Bethlehem, Jericho, Hebron, and Nablus. It belonged to Palestinians until the third Arab-Israeli War. In 1993 an agreement was reached allowing the region to exist as a Palestinian enclave. The Gaza Strip is another disputed area. It was part of the Egyptian Sinai after the first Arab-Israeli War until 1967, when Israel occupied it until 1994. Today the Gaza Strip represents a Palestinian enclave. These regions have changed hands throughout successive wars.

The first Arab-Israeli War took place in 1948 in response to the establishment of the state of Israel, when Egypt sent forces into Palestine on two fronts. The Israelis sprang a surprise attack, defeating the Egyptians. The second war occurred in 1956, when Israel invaded Egypt's Sinai Peninsula. The third happened in June 1967 and lasted six days. The war began with Egypt blocking the Tiran Straits, paralyzing Israeli ships. At the same time, Iraqi forces moved into Jordan, ready to attack. Israeli forces acted swiftly and captured East Jerusalem, the West Bank, Golan Heights, and Sinai (which they returned to Egypt in 1979). The fourth war had a much larger global impact because conflict over the region resulted in the 1973 oil embargo. Because of this global dimension, the war has received three other names based on the political stance of onlookers. By supporters of Israel the war is referred to as the Yom Kippur War because it took place on the holiest day on the Jewish calendar. It is also referred to as the Ramadan War by supporters of Palestine because it took place during the Islamic holy month. By others, it is referred to simply as the October War because of when it occurred on the standard Gregorian calendar. The war broke out when Egypt and Syria invaded Israel and demanded the return of borders to their pre-1967 locations. A cease-fire was reached after three weeks of fighting. The fifth, and final, war began in 1982 when Israel invaded Lebanon. The situation in the region has not been resolved. The Palestinians and the Israelis, however, have moved toward a series of agreements establishing mutual recognition and a possible end the long-standing conflict.

Having placed the Arab-Israeli conflict in historical perspective, let us go back to the 1973 oil embargo. The objective of the embargo was twofold: impact both Israel directly and its largest supporters in the Western states. First, the Arab states planned to prevent oil from reaching Israel as a way of creating dissatisfaction among the Israeli people who, in turn, would challenge their government. Second, the Arab states planned to force the countries of the West, primarily the United States, to choose between their ties to Israel and their need for oil. The Arab states assumed that the countries' economic needs would outweigh their emotional attachment to Israel. The oil embargo officially began in October 1973 and ended in March 1974. Essentially, OAPEC members decided to refrain from sending oil to countries that provided financial or military support to Israel. These countries included the United States, primarily, and select countries in Europe. OAPEC saw that the world demand for oil was high and that a decrease in production from any two Arab countries could cause a panic among consumers in the West. The main goal of the embargo was simply to attract public attention in the West to the Palestinian-Israeli question.

On October 6, 1973, the fourth Arab-Israeli War, or October War, began and lasted until the 22nd. During the war, ministers from the Middle East asked the United States to withdraw its support for Israel. The United States, instead, increased its support prompting the Arab nations to declare that oil shipments to the United States would stop until the Israelis agreed to pull back to their 1967 borders (the ones in force before the third Arab-Israeli War) within a given period and with U.S. support. The United States and the Arab states were now locked in opposition over the political turmoil in the Middle East, and the Arab countries began to forge a new sense of unity. On October 17, the flow of oil to the United States ceased.

Although they generally did not know it, consumers in the United States depended on Middle Eastern oil a great deal. The United States produced less and less oil for domestic consumption every year and imported more and more from overseas sources. The major oil companies attempted to shift their oil supplies away from the Middle East. Until they could effectively do so, they raised the price in order to counter the risk of remaining in the politically unstable region. Observers argued that the companies used the political crisis in the Middle East as an excuse to raise prices. No matter how much the major oil companies did actually tinker with the price, not having the world's largest oil producers shipping oil to the United States created a shortage of readily available oil and drove up the price of petroleum products throughout the world. Over a span of a few months, the price of oil quadrupled. In Japan the price of oil increased three times as much as in the United States from 1973 to 1974. In Western Europe, the impact was less dramatic partially because governments such as the United Kingdom and France made special deals with oil-producing countries in the Middle East, securing the continual flow of oil. Also, the United Kingdom supported the Arab demand for Israel to withdraw to its pre–1967 borders and several European countries, including the Netherlands, made a collective statement seen as pro-Arab regarding the Arab-Israeli conflict. The greatest impact on Western Europe was the fear of a shortage and the uncertainty of when the embargo would end. But Western Europe was not seen as the primary target. It was the United States who provided the most support to Israel and consumed a great deal of oil from the Middle East.

During the embargo, the United States faced severe gasoline shortages and rising unemployment as the transportation industry contracted. The popular news magazine *Time* documented the impact of the 1973 oil embargo on U.S. consumers, citing the changes that took place:

- The reign of the big, gas-gulping car in the United States ended abruptly, as consumers began to buy smaller, more fuel-efficient cars.
- The speed limit on major highways was reduced from 70 mph to 55 mph.
- Several states on the East Coast adopted rationing plans to avoid lines at filling stations and curb overconsumption.
- Independent truck owners took to the highways to protest the rise in price and scarcity of diesel fuel.[6]

While U.S. citizens braced themselves for a long, hard winter in 1973, the oil companies weathered hostilities from all sides. Consumers, and even the U.S. government, blamed the oil companies for the high prices and for creating bad blood with the Arab oil producers. The U.S. government even went so far as to conduct a hearing on the oil companies' involvement and to consider nationalizing Exxon. At the same time, the oil companies faced pressure from the oil-importing countries to solve the problem. The oil companies, of course, also feared for the future of their markets and profits. Not only did they have to solve the immediate problems of shortages, but they also had to find a strategy to prevent future crises. Fortunately for them, they did not have to agonize very long.

By the end of the year, the Arab states had withdrawn their restrictions on Europe, and a few months later, in March 1974, they lifted the embargo on the United States. One year later, the world supply of oil met demand and OPEC ensured that the price of oil went down and stayed at a fair price. As quickly as the embargo had created a panic within the United States and Europe, it subsided. The United States vowed to become self-sufficient in energy by 1980 through expanding its domestic oil production and looking into nuclear power. This plan did not materialize, however, and the United States maintained its dependence on the Middle East (even to this day). Overall, the impact of the embargo was traumatic but not as crippling as the OAPEC members had hoped. This was because the embargo lasted less than one year and because the oil industry is so complex that it is virtually impossible for a producing country to control where its oil actually goes once it is pumped out of the ground. In fact, the Arab states should have known this; 1973 was not their first attempt at an oil embargo.

The 1973 oil embargo was not the first attempt by Arab states to implement an oil embargo as a political weapon. During the third Arab-Israeli War, in 1967, the belief spread among the Arab League that Israel was supported by Britain and the United States.[7] Arab leaders met in Baghdad and agreed to shut down the oil wells and to boycott the West. The boycott was short-lived because it was ineffective, particularly against the United States. Furthermore, the boycott actually increased profits for non-Arab oil producers such as Venezuela. The 1973 embargo represents just one example of oil used as a political weapon within international politics. In Chapter 5 we will return to this issue in regard to how a country's or organization's foreign policy affects the world oil market.

As in the attempted boycott of 1967, oil-producing countries in general benefited from the 1973 embargo. Non-Arab oil-producing countries such as Iran, Mexico, Nigeria, Norway, Venezuela, and Russia enjoyed the influx of oil revenues that the embargo brought without any of the political repercussions of taking part in the embargo itself. The high oil prices during the embargo made new exploration and production projects much more economically feasible. For example, the high oil revenues made offshore drilling—an expensive way of producing oil—much more profitable for Norway. The minor oil companies also benefited from the embargo. The major oil companies, having been paralyzed, feared that the smaller oil companies producing outside of the Middle East would step into the vacuum and charge outrageous prices for their petroleum products. This, in fact, was the case all over the world, including in the United States. In the end, everyone benefited from the oil embargo but consumers in the United States.

The two events together—the price revolution and the embargo—had a dramatic impact on the world's oil industry. The combination of events, however, came about largely by chance. The oil embargo was meant to attract interest among the Western public to the Israeli question, while the oil price revolution was an attempt by oil producers to increase their power as well as the price of oil. Regardless of the fine distinctions between events, the Western response was to protect itself from the first oil shock as well as potential oil crises in the future. Just as the uncertainty of the 1960s brought unity among oil-producing countries in the form of OPEC, it also pulled the United States and Europe closer together and prompted them to create the Organization for Economic Cooperation and Development (OECD).

ORGANIZATION FOR ECONOMIC COOPERATION AND DEVELOPMENT

The OECD, formed in 1961 to foster cooperation, economic growth, and trade. The organization is currently made up of thirty countries (Figure 4.3). Perhaps influenced by the Cold War, the members expressed their commitment to the promotion of democracy and a free-market economy. They also pledged to preserve human rights. The formation of the OECD was described as the Western response to the formation of OPEC in an effort to renew the solidarity between the United States and Western Europe, which had weakened since World War II. On December 14, 1960, twenty countries agreed to form the organization, with several new members invited to join periodically. According to one scholar, the major weakness to the organization has always been a raison d'être, because it had only a few commitments that governments were pledged to carry out.[8] The most evident point of unity and focus for the organization came during the 1973 oil embargo.

Figure 4.3
OECD Members

Australia (1971)	Korea (1996)
Austria (1961)	Luxembourg (1961)
Belgium (1961)	Mexico (1994)
Canada (1961)	Netherlands (1961)
Czech Republic (1995)	New Zealand (1973)
Denmark (1961)	Norway (1961)
Finland (1969)	Poland (1996)
France (1961)	Portugal (1961)
Germany (1961)	Slovak Republic (2000)
Greece (1961)	Spain (1961)
Hungary (1996)	Sweden (1961)
Iceland (1961)	Switzerland (1961)
Ireland (1961)	Turkey (1961)
Italy (1962)	United Kingdom (1961)
Japan (1964)	United States (1961)

In the face of rising oil prices and political turmoil within the Middle East—the heart of the world's oil—OECD strengthened its membership during the early 1970s. The organization focused its efforts on how the advanced countries could establish greater economic independence. As the embargo's major target, the United States encouraged the OECD members to help one another triumph over any problems stemming from an increase in oil prices and set up conservation and sharing agreements to ease their reliance on OPEC as a source of petroleum. The advanced countries worked as one in opposition to OPEC; the greater their bargaining position in the oil market, the more likely they would be able to drive down the high prices for oil. In 1974 the United States led sixteen advanced countries in the formation of the International Energy Agency (IEA) in Paris, which sought to improve the response among member countries in the face of oil supply shortages. During the oil embargo the IEA promoted collaborative work in securing supplies of oil for its members and encouraged members to set up a ninety-day reserve of crude oil. The IEA continues to operate today and provides valuable statistical data on energy consumption and production around the world. In addition to the years of data they have amassed, they have also received a great deal of criticism.

The OECD has received criticism for contributing to the further economic disparity between developed countries and developing countries. It is also accused of encouraging paternalism among rich countries as a method of self-protection.[9] Perhaps, however, the OECD's goals require reexamination now that Mexico has joined the organization. Mexico, which joined the OECD in 1994, represents the first Latin American developing country—and one of the world's largest oil exporters—in the organization.

AN INDUSTRY DIVIDED?

In this chapter, we introduced three major organizations active in the oil industry—OPEC, OAPEC, and OECD—highlighting the motivations for unifying along ethnic, political, and economic lines. Although the 1960s and 1970s could be described as two decades of unification, on closer examination we see that the creation of these organizations has resulted in a more divided world, in which oil plays a prominent role. OPEC formed because its members sought to implement a collective strategy to gain more control of their industry and increase their returns from the oil profits made by the foreign oil companies. Then, several Middle Eastern countries formed OAPEC to promote the idea of depoliticizing oil and reinforcing the bonds among Arab oil-producing countries. In response to OPEC and OAPEC, oil-importing countries with similar economic standing within the West formed the OECD in an effort to renew solidarity and protect its members from potential fuel crises. Through the actions of these groups, the separation between the oil business and foreign politics became blurred. In the 1970s two major events marking the high point of oil politics occurred simultaneously: the oil price revolution and the 1973 oil embargo. Oil became more than a valued commodity on the market; it became a powerful political weapon. In the following chapter, we will look at the use of oil not only as a weapon, but also as an olive branch between companies and countries all over the world.

FURTHER READING

OPEC has been the center of international power and intrigue since its formation. A terrific starting point is OPEC's Web site (www.opec.org), which provides up-to-date information on its activities and members as well as an overview of the organization's history. Its activities have been a major topic of research for scholars, with many of their works providing a solid overview of the organization. These works include Mohammed E. Ahrari's *OPEC: The Failing Giant* (Lexington: University Press of Kentucky, 1986), Fu'ad Ruhani's *A History of O.P.E.C.* (New York: Praeger, 1971), and Jahangir Amuzegar's *Managing the Oil Wealth: OPEC's Windfalls and Pitfalls* (New York: I. B. Tauris, 2001).

A wealth of information on the 1973 oil embargo exists. A few works, however, stand out as engaging and readable. See Anthony Sampson's *Seven Sisters: The Great Oil Companies and the World They Shaped* (New York: Viking Press, 1975), and Daniel Yergin's *The Prize: The Epic Quest for Oil, Money, and Power* (New York: Simon and Schuster, 1993). Jeffrey Robinson's *Yamani: The Inside Story* (London: Simon and Schuster, 1988) tells the life of Ahmad Zaki Yamani.

For information on OAPEC, see Abdelkader Maachou's *OAPEC: An International Organization for Economic Cooperation and an Instrument for Regional Integration* (New York: St. Martin's Press, 1983), and OAPEC's Web site (www.oapecorg.org), which provides the history, membership, and goals of the organization.

For more on the Israeli-Palestinian conflict, see Charles D. Smith's *Palestine and the Arab-Israeli Conflict* (New York: St. Martin's Press, 1988). For excellent overviews on Islam, see Thomas W. Lippman's *Understanding Islam: An Introduction to the Muslim World* (New York: New American Library, 1982), and on Judaism see Nicholas de Lange's *An Introduction to Judaism* (New York: Cambridge University Press, 2000).

For information on the OECD, see the OECD's *The OECD: History, Aims, Structure* (Paris: OECD, 1960), and Miriam Camp's *"First World" Relationships: The Role of the OECD* (New York: Council of Foreign Relations, 1975). Also see its Web site (www.oecd.org) for up-to-date energy statistics and information about member countries.

Part II

THE ISSUES

Chapter 5

International Politics

Oil has a key role in the world economy and we cannot rely on market forces alone to provide stability and prosperity that we all depend upon.
Rilwanu Lukman, 1998[1]

From the moment of its discovery, oil quickly became a major commodity traded and used in building nations both economically and politically. During the first half of the twentieth century, petroleum aided nations who had access to it, giving them an advantage during military operations over nations who did not. Much of the success of the Allies during World War II can be attributed to their access to oil and their ability to cut oil supply lines to the Axis powers. Today, not much has changed. International politics has evolved to the point where there is hardly a corner of the world where oil does not play a prominent role.

Oil acts as a stimulus that pushes or pulls countries in often strange, new ways. While the thesis is difficult to prove, it is quite possible that without the allure of petroleum, countries would not have adopted the political agendas that they did. Perhaps their positions as players in international politics would have been quite different. The desire or need for oil has pushed countries into conflict: regional or civil war, territorial disputes, and economic sanctions. At the same time, oil pulls countries into international politics. This is particularly true of oil-producing countries. Oil encourages countries to share their valuable resource and the wealth they derive from it. Oil-rich countries have extended economic assistance to other countries through trade, aid, regional alliances, and joint development projects. While there are numerous examples of each, those included here are among the most successful and the most destructive. In many ways these countries represent pioneers within the global community for their positive and negative contributions to the geo-political landscape. First let us look at the less publicized positive efforts before moving on to the disastrous circumstances where oil has led to war.

INTERNATIONAL COMMUNITY BUILDING

During the 1960s, blocs of nations unified into international organizations centered on petroleum. For oil-producing countries, OPEC represented only the beginning of alliance building. While the members of OPEC structured a cooperative relationship with producers on faraway continents, they also took their relationship building to the regional level. This was especially true during the first and second oil shocks, when oil prices on the international market skyrocketed and oil-producing countries everywhere enjoyed an influx of unprecedented oil wealth. Developing countries, particularly, engaged in a new foreign policy of assisting other developing countries. The oil-rich began to help the oil-deprived.

With a newfound sense of prosperity and economic security, oil-rich countries saw that one of the best ways to assist neighboring countries during the oil shocks was by helping them handle the high cost of oil imports. Oil-rich countries did this in two ways: they provided direct aid and they sold oil at discounted prices. Venezuela, for example, used a portion of an investment fund during the 1970s to assist countries in Central America and the Caribbean in affording the high cost of imported oil. The Fondo de Inversiones de Venezuela (FIV) was created in June 1974 to promote state investment within and outside Venezuela. The FIV project has been criticized to promoting government spending on luxury items in Venezuela more than encouraging regional aid and internal development. President Andrés Pérez, however, received praise for using the fund to contribute millions of dollars in regional aid. Through the fund, Venezuela contributed $100 million to the United Nations Emergency Program and $224 million to OPEC's Special Aid Fund in 1977.[2] In 1980 Venezuela and Mexico began supplying crude oil and refined petroleum products to eleven Central American and Caribbean countries through the San José Accord.

The San José Accord, a trade arrangement, worked like this: The oil-producing country set aside portions of its crude oil to send to a relatively nearby oil-importing country. The importing country purchased this crude oil at a reduced price and then refined and distributed it. Venezuela and Mexico supplied up to 80,000 barrels of crude oil per day each. Through this agreement, Mexico and Venezuela increased their involvement in the affairs of Central America and the Caribbean. The San José Accord was designed to help economically weak nations weather the potentially destabilizing high costs of petroleum products. While the objectives among these countries were philanthropic in design, other forms of community building have a different motivation.

In an effort to promote regional trade and ease the flow of petroleum products among countries, the European Economic Area (EEA) agreement as well as the North American Free Trade Agreement (NAFTA) were signed during the 1990s. Together the countries in North America and Western Europe, the two dominant economic regions of the world, strengthened regional alliances and improved the exchange of goods such as natural gas and foodstuffs.

The membership of each alliance remained small, with fewer than ten nations in total. In 1992 the EEA formed under the umbrella of the European Union. Two years later NAFTA was formed among Canada, Mexico, and the United States. Currently NAFTA represents the world's largest free trade area, while the EEA includes many of the world's largest companies. Within each agreement, import restrictions of trade are loosened and standards for labor and environmental preservation tightened. In other words, the tariffs and taxes that protect vulnerable sectors of a country's economy are lifted within the alliance but remain in place for a country outside the alliance. While this arrangement works well for developed countries such as those in Europe and the United States, it poses a problem for Mexico's fragile agricultural sector. Both the EEA and NAFTA have encouraged the use and improved the trade of natural gas within their respective regions. NAFTA has attempted to turn Mexico from an importer into an exporter of natural gas. One way it has assisted Mexico in doing this is through working around Mexico's state oil company, PEMEX, and opening the country to private companies that can properly invest in Mexico's natural gas industry. By joining NAFTA, Mexico locked in access to the region's largest market—the United States—and restored confidence among private investors in its government. Through these regional alliances, oil- and gas-producing countries such as Mexico and Norway have secured a reliable market for their natural gas. Similar relationships have emerged regarding the development of cross-country pipelines.

Within the past few years, two major cross-country pipelines have been in the works. The Caspian Sea Pipeline Consortium went onstream in

2001 and connects Kazakhstan's Caspian Sea oil fields with Russia's Black Sea port of Novorossiysk. The Caspian Sea Consortium is unique in that six countries—Azerbaijan, Iran, Kazakhstan, Russia, Turkmenistan, and Uzbekistan—claim a portion of the pipeline. While territorial claims of the Caspian Sea itself remain somewhat unclear, Russia went ahead and secured its oil interests through a pipeline. The Caspian Sea represents an up-and-coming oil and gas location with reserves estimated to contribute around 4 percent of the world's total. The project was funded primarily by Russia and, no doubt, serves its interests as a major oil power within the former Soviet Union. Likewise, a cross-country pipeline is under construction in West Africa. The West African Gas Pipeline will travel over 600 miles from Nigeria's Niger Delta to a terminal in Ghana. In 2005, ChevronTexaco announced that construction would begin in December 2006, and the pipeline would go onstream in 2008. The project has been primarily funded by ChevronTexaco in Nigeria and the NNPC. Through this project, Nigeria is providing natural gas to its oil-importing West African neighbors, which serves to reinforce its dominant role in the region.

Projects connecting oil producer and oil importer have occurred in unlikely ways. Regional cooperation generally works based on geographic proximity and often a mutual political and cultural background. Frequently a common language, history, and culture already connect the two, making the economic connection a logical progression. In some cases, however, alliances are forged in less conventional ways. For example, within the past few years Venezuela and Zimbabwe have developed plans for a joint-venture project. In 2004 President Hugo Chávez of Venezuela and President Robert Mugabe of Zimbabwe signed an energy cooperation agreement, which called for assistance in exploration and sales. This agreement represents one of the few examples of formal economic cooperation between African and South American countries. Zimbabwe called on Venezuela as a way for developing nations to help each other instead of relying on major oil companies from the United States or Europe. Zimbabwe periodically suffers from oil shortages and hopes to relieve this problem; it also faces extreme social and political conflict.[3] What brings these two seemingly different countries together? The answer lies in a closer look at each of the country's leaders.

Mugabe and Chávez both made their names as populist leaders with a strong authoritarian streak. Chávez became a major political figure in Venezuela through a failed military coup in 1992, followed by a landslide win in a 1998 presidential election. Mugabe became prime minister of Zimbabwe in 1980 and has remained in power ever since, through questionable popular elections. Both men have maintained an interest in agrarian reform that could improve the living situation for the poor, Zimbabwe's with marked racial overtones. Both leaders view the United States with great skepticism because of its handling of the situation in Iraq. Also, both have been in positions of power for almost twenty years, and in the twenty-first century, their popularity is waning both within the country and in the international community as a whole. For these reasons, the two leaders have sought political and, now, economic support from each other. But is this relationship—and others like it—about cooperation, or are there sinister motives behind it?

During the cold war, relationships like these appear to have been straightforward cases of self-interested politics. After World War II, the United States and the Soviet Union expanded their political alliances throughout the world. When the British and other European empires began to erode in the 1950s, the Soviet Union and the United States rushed to the newly independent countries throughout the Middle East, Asia, and Africa to secure their support against the other. Many of these countries were oil producers that played one major power against the other. Each country found a way to use the tension between the United States and the Soviet Union to its advantage. The Soviet Union used one of its greatest assets—oil—to sway developing countries, while the United States used its major oil companies to do the same. The Soviet Union sold its oil at reduced rates to the developing countries that needed to import oil, while it sent experts to oil-producing countries to assist them in developing their own oil industries. The United States sent in their oil companies not only to assist in developing an oil industry, but also to invest in the public sector (i.e., roads, health services). While the cold war is more commonly known for its "arms race," there was also an "oil race" to help oil-producing countries develop their industries as well as to compete for oil-importing markets. The cold war is the perfect example of two major powers using oil to fulfill their own political agendas and the newly independent states taking what advantage they could of the situation.

Philanthropy or Power?

All these activities—from building cross-country pipelines to providing aid—can be interpreted in two ways. First, they can be viewed positively as steps toward mutual cooperation among oil-rich and oil-poor countries. Yet the behavior of a nation is inherently based on self-interest. It is safe to say, then, that no country would enter an economic relationship without gaining something. With this in mind, these relationships can be viewed as yet another forum for gaining political and economic power. The evidence lies in the outcomes of these agreements and contracts. For example, aid projects such as the San José Accord took place mainly during the first and second oil shocks, when oil prices on the international market skyrocketed and oil-producing countries everywhere enjoyed an influx of unprecedented oil wealth. The end result of the accord was to turn many oil-importing countries into debtor countries. Shipments of oil to Nicaragua were suspended in 1982 when it failed to pay a debt of $18 million to Venezuela and $300 million to Mexico for oil it had previously received.[4] The attempt to balance power among oil-rich and oil-poor countries turned into a strained loaner-debtor relationship. The same sort of harmonious, cooperative beginning also marked the building of cross-country pipelines.

Like any huge joint project, an oil pipeline takes a great deal of capital, commitment, and patience. In some cases, it eventually creates tension among neighbors. In the case of the pipeline running through West Africa, any problems with the pipeline fall on the shoulders of Nigeria because of its heavy investment and involvement in the project and because it has a long-standing history as the dominant political force within West Africa. As the pipeline project got under way, complaints about environmental destruction and population displacement emerged, with the blame falling on Nigeria. In the case of the Caspian Sea pipeline project, Russia plays the dominant role because it has contributed the most capital to the project. Regional instability has made building the pipeline difficult. Also, territorial claims over the Caspian Sea have not been completely resolved, which has made the Caspian pipeline project more a source of conflict than cooperation.

ARRIVAL OF OIL CONFLICT

The drive to acquire oil has pushed countries into conflict in the form of war and territorial disputes. While these conflicts are certainly not new, they have been increasing over the years. Since its discovery in the nineteenth century, oil has played a prominent role in political confrontations. During World Wars I and II, oil assisted in determining who was victorious. In World War II, for example, the Allies' success can be attributed to their access to Middle Eastern oil and their ability to cut off the Germans from their oil supplies. Today political power is determined by a country's access to oil in both similar and new ways. Since World War II, conflicts have emerged over oil itself, instead of oil's just playing an important role in military campaigns. This escalation of oil-related conflict can be attributed to an increase in global demand for petroleum products that drives countries to secure new sources of oil.

The expansion of worldwide demand has resulted in countries scrambling to find new sources of oil and natural gas. The world's population has gone from 2.5 billion in 1950 to 6.5 billion in 2004. This rise in population has increased the world's need for food and for everything else that satisfies basic living requirements. Not only does oil serve as a source of fuel, but almost everything we use in our daily activities. Almost everything from clothing to plastics to fertilizers is a petroleum derivative. Humans are extremely dependent on petroleum products, and current consumption trends indicate that this dependency will only increase. As the value of petroleum increases, global conflict over the resource has intensified.

Oil-related conflict manifests itself domestically and internationally, violently and peacefully. In most cases, the conflicts are resolved through negotiation and without violence. Some conflicts, however, result in violence, with high casualties on both sides. Whether an oil conflict becomes violent or not really depends on the economic and political situation of the parties involved. In some cases, the formal judicial process is not successful, for example, in the context of a border dispute between neighboring countries, or is not viewed as a viable option, when, for example, minority groups within a country lack support from the central government and must resort to violence. These conflicts usually result in civil wars, as in the cases of Nigeria and Russia. Resorting to violence has a great deal to do with how desperate a country's situation has become. Countries that view the acquisition of oil as essential to their survival will take the high risk of war, whereas a country that already has enormous oil wealth will not take such a drastic move. To properly flesh out the various types of oil conflict, let us first turn to conflict that takes place within an oil-rich country's borders.

Domestic Conflict

While the focus of this chapter is on the relationships among nations, it is important to explore briefly the role of oil in domestic politics. In most oil-rich countries, of course, oil is not evenly distributed geographically so that all corners of the country benefit from the oil industry (or suffer from environmental damage created by it). For most countries in the world, economic development is not evenly dispersed and adding oil to the already lopsided picture creates more problems. A developing country struggling to establish national unity finds the location of oil an on-going problem. In several cases of oil-rich countries, the regions rich with oil do not want to remain part of the nation; they want to secede and establish independence. Much to their dismay, the national government has no interest in letting them go free without a serious fight, because both the government and the national economy are heavily dependent on the oil industry. In the cases of Nigeria and Russia, this political struggle has erupted into civil war.

Nigeria's civil war from 1967 to 1970 was the result of an ongoing conflict among competing regional powers and ethnic groups. The eastern region (including the oil-rich Niger Delta) seceded from Nigeria to form the Republic of Biafra. The secession had an enormous impact on Nigeria because the eastern region took much of Nigeria's oil industry with it. By 1970, the Republic of Biafra could no longer sustain its independence and rejoined Nigeria. Russia faced a similar problem with Chechnya. The struggle between the Chechens and the Russians extends back to the nineteenth century, when the Chechens demanded permanent independence. The Russian government tolerated their resistance movement for several years, but the importance of oil pipelines and railway links prompted Russia to secure its claim over the region by force. Chechnya is located on the Caspian Sea, one of the major offshore oil-producing regions in the world. Inhabited by primarily Turkic Muslims in the Caucasus Mountains, Chechnya became an independent republic in 1991. In 1994 Russian armed forces stormed Chechnya and destroyed Grozny, the capital. Five years later, the Muslim Chechens attacked the republic of Dagestan. Moscow responded with brutal force to stop the incursion. In 2000, Russia reclaimed Chechnya by declaring direct rule. If Chechnya became independent, Russia would lose its claim to a portion of the Caspian Sea's oil and gas deposits. The presence of the pipeline from the Baku oil fields on the Caspian Sea through Chechnya to the Russian port of Novorossiysk on the Black Sea is vital to Russia. In both countries, the national desire to keep the oil-rich region within the country has been the country's driving force for civil war. Without the promise of oil, perhaps the war could have been avoided. The same can be said in the case of territorial disputes revolving around existing or potential oil and natural gas reserves.

International Conflict

Onshore Disputes

Territorial claims often surface when the modern borders of a country were determined by foreign colonial rulers, such as the Spanish, British, and French. Many of the oil-rich countries today spent previous decades under foreign occupation. Their former colonizers determined where the expanse of one country began and another ended, regardless of the location of the region's people. As natural resources, especially oil, became important to a country's economy, borders became particularly important. Many of the oil fields in the world are in areas where national borders have not yet been clearly drawn. This is especially true where borders were originally determined by geographic barriers such as rivers or mountains. This situation occurs primarily in the tropical regions of South America. Venezuela claims the Essequibo River region west of its neighbor Guyana. The region is also a favorite tourist destination because of its lush tropical terrain. The region is rich in oil as well as gold and diamonds. In 2004, Venezuela invited foreign oil companies to drill for oil and gas in the region, inflaming a territorial dispute with Guyana that dates back to the nineteenth century.[5] This mutual claim demanded the intervention of the United Nations and has yet to be resolved.

In the case of the Middle East, borders are delineated largely by lines of longitude and latitude running through uninhabited desert. For this reason, Saudi Arabia has several border disputes. Because of the Arabian Peninsula's vast expanse of desert, territorial claims for large tracts of sand did not occur until there was potential oil underground, and, of course, Saudi Arabia is now the world's largest oil producer. While most countries have clearly defined borders that were physically marked by fences and border patrols centuries ago, countries on the Arabian Peninsula are just now firmly defining their borders. Saudi Arabia shares a border with Iraq, Kuwait, Oman, Yemen, Qatar, and the United Arab Emirates. Most of the border disputes involving Saudi Arabia have existed since the end of British colonial rule. Saudi Arabia and Yemen began a dispute over the location of a mutual border in the 1930s. In 1994, Saudi Arabia and Yemen fought over the location of the border, and in 2000 the two countries finalized it. Plans have been in the works for possible oil and gas exploration as well as a cross-border pipeline. A couple of years later, Saudi Arabia clashed with the United Arab Emirates over the location of their mutual borders. The dispute with the United Arab Emirates occurred when in 1998 Saudi Arabia began pumping oil from an oil field that straddles a disputed border between the two countries. Once oil production began, the United Arab Emirates, naturally, demanded a share in production. The Saudi Arabian and Venezuelan cases occurred where the borders were not clearly defined, but in some instances, oil-thirsty countries move forward across clearly defined borders intentionally.

In two instances, Saddam Hussein attempted to expand Iraq's control over its regional neighbors in a ruthless quest for more oil. Both the Iran-Iraq War (1980–1988) and the Iraqi invasion of Kuwait (1990–1991) represented blatant acts of greed. Saddam Hussein came to power in 1979 and ruled despotically until March 2003. The Iran-Iraq War broke out one year after he came to power due to a territorial dispute over who controlled the Shatt-el-Arab waterway. Competition for regional dominance as well as political and religious differences had set the two countries at odds for many years. When Iraq invaded Iran that year, Iraq envisioned the invasion to be short and decisive, because Iran was already severely weakened from its 1979 revolution. In reality, the opposite happened; the war dragged on until 1988. After eight years of fighting, the war ended in a stalemate, with neither side advancing politically or territorially. Iraq suffered high casualties and accrued war debt. None of this, however, prevented Saddam Hussein from launching another invasion a little more than a year later.

Shortly after the Iran-Iraq War, Saddam Hussein once again involved Iraq in a territorial dispute that led to war. Iraq's claim to territory belonging to Kuwait is not new. Within days after Kuwait received its independence from Britain in 1961, Iraq sought to annex the country. The Arab League and British troops rushed to Kuwait's aid, and Iraq chose not to press its claim at that time. But in 1984 Iraq under Saddam Hussein pressed its claim on Kuwait to cede two islands (Warba and Bubiyan) to Iraq. Naturally, Kuwait refused and reinforced its border security.

Iraq's invasion of Kuwait can really only be explained by its desire for oil. Saddam Hussein embraced a peculiar form of Pan-Arabism, which included a belief that wealth and resources should be shared among Arab states. Furthermore, he contended that Kuwait only existed as an artificial creation during British rule and that it really belonged to Iraq. Kuwait, according to him, should not exist as a separate state. He argued that the small state of Kuwait should share its oil wealth more evenly among Arabs. The timing of this claim stems from Iraq being burdened by external debts to Kuwait, among others. Finally, Iraq also needed a greater water frontage on the Persian Gulf, which seizing Kuwait would provide.

In August 1990 Saddam Hussein continued his designs on Kuwait by ordering a full-scale invasion to seize control of its rich oil fields. Shortly after the occupation, the United Nations intervened. Iraq responded by reinforcing its presence in the small country. Because this occurred along the Saudi Arabian–Iraqi–Kuwaiti mutual border, Saudi Arabia called for U.S. protection. The United States swiftly responded by moving security forces in to protect the country from possible Iraqi advances. The United Nations demanded that Iraq leave Kuwait and gave the country a deadline for withdrawing. Seven months later, under international military and political pressure, Iraq withdrew from Kuwait. In the end, Iraq did not increase its control over oil fields and, in fact, it lost a portion of its own fields to wartime destruction.

While the United Nations has played a major role in dealing with the problems in Iraq, it does not always assume this role. The United Nations is not the ultimate arbiter of international action toward a harmful country. Any move by the United Nations requires a consensus among its members. Failure to reach a multilateral decision does not prevent individual members from moving forward with military aggression. The United States, for example, moved ahead with its plans to remove Saddam Hussein from power without the support of the United Nations in 2003. Without confirmed evidence regarding Iraq's weapons of mass destruction, the United Nations did not want to invade. In March 2003, the United States, backed by Britain, gave Saddam Hussein and his sons forty-eight hours to leave Iraq or face war. On March 20, 2003, the United States invaded Iraq. The United Nations did not explicitly support this U.S.-led campaign, but it maintained an active interest in the future of Iraq after Saddam Hussein was effectively removed from power. The United States, however, set up the interim government and made all the major decisions regarding the future of Iraq. Critics have argued that the United Nations should have taken a more active role.

The United States, and more specifically George W. Bush's administration, has received a great deal of criticism for having more oil-related than humanistic reasons in mind. According to Bush's critics, Saddam Hussein represented a greater threat to the interests of U.S. oil companies than to the world. The current U.S. administration is comprised of people who gained their millionaire wealth through oil. Vice President Dick Cheney made his millions when he served as the chief executive of Haliburton, an oil services company, which received a contract to rebuild Iraq's oil industry after the 2003 war. While Cheney denies any connection to the company, the public has been suspicious. The removal of Saddam Hussein has left many production contracts between the Iraqi government and major oil companies uncertain. Hussein had reportedly signed several multibillion-dollar deals with foreign oil companies, mainly from China, France, and Russia. In 2003, Philip Carroll, the chairman of the U.S. committee on the planned management team for Iraq's oil industry, stated that contracts signed under the previous regime would be assessed to determine whether they were made with the Iraqi people in mind. Critics saw this as an opportunity for the Bush administration to handpick the oil companies to be involved. Nations that opposed the war, for example, are not allowed to bid on contracts. Thus, France, Russia, and Germany, in particular, have lost their presence in Iraq. Canada was also part of this group, but the United States has reversed its policy and may allow it to bid in the second round.[6] Prior to taking on the position of chairman of the U.S. committee on the planned management team for Iraq's oil industry, Carroll acted as chief executive of the U.S. division of Royal Dutch/Shell. Finally, the United States has not provided sufficient evidence of weapons of mass destruction or clear connections to Osama bin Laden and al-Qaeda, which were the reasons used for the invasion. While no one denies the benefits of removing Saddam Hussein from power, the tie between the Bush administration and Iraq's oil has raised suspicion.

Offshore Disputes

The importance for oil-producing countries of being located next to or having access to large bodies of water has increased because having a coastal outlet provides an oil-producing country with an easy place from which to export its petroleum products. Also, within the past twenty years, being located next to water has meant a potentially new source of oil and natural gas reserves. For decades, oil companies thought in terms of oil exploration only on land. Offshore oil production was not economically or technically feasible for many years. For this reason, Norway's oil industry did not develop until the technology for offshore oil production was in place. Offshore oil and natural gas production has provided an entirely new avenue for oil-producing countries that face possible decline from their onshore oil fields. For these reasons, access to water has become an increasing source of international conflict.

Territorial disputes become more complicated when offshore oil production is involved. The tensions that exist between neighboring countries usually involve a disagreement over who legally claims a continental shelf that extends into a mutually bordered body of water. The development of international law pertaining to national sea rights is recent. Prior to the formation of the United Nations Convention on the Law of the Sea (UN-CLOS) in 1958, national rights extended only to the coastline, which generally meant a little over 3 miles from the coast. The remainder of the water was *international waters*, free to all nations but not owned by any one nation. Due to fishing and oil exploration interests, countries violated this long-standing informal agreement. The United States led the way in 1945 by establishing that its borders extended out 200 nautical miles, or approximately 230 distance miles from its coastline.

Under the auspices of the United Nations, an international conference took place in 1958 to consider national sea rights. After three conferences, UNCLOS signed a treaty in 1973 establishing territorial claims in oceans which went into effect in 1994. The treaty established the limit of aquatic territory, where a nation had complete sovereignty, at 12 nautical miles (13.75 distance miles). Beyond the twelfth nautical mile, a "contiguous zone" existed where a state could enforce laws against illegal activities such as smuggling. In the area up to 200 nautical miles from shore, the exclusive economic zone (EEZ), a nation had exclusive rights. While these laws determined how each zone could be used, they did not clarify where borders existed.

In the 1970s, a political clash between Nigeria and Cameroon was triggered by mutual claims to offshore oil reserves along the southern portion of the common border known as the Bakassi Peninsula located in the Gulf of Guinea. Cameroon and Nigeria both claimed the 400-square-mile area, believed to contain large oil reserves. The dispute led to clashes and the detainment of prisoners of war on both sides. After several diplomatic negotiations, an accord was signed in 1974. In 1994, the case went to the International Court of Justice. Two years later, Nigerian forces clashed with Cameroonian police in the disputed region. In 1998, formal hearings took place with the two countries exchanging prisoners of war as a gesture of peace. The case became further complicated with Equatorial Guinea's claim to part of the water rights. In 2002, the court awarded the Bakassi Peninsula to Cameroon. Unfortunately, international law has not been as helpful in the case of the Caspian Sea.

The Caspian Sea is considered the next major oil- and natural-gas-producing region in the world. With this potential, ownership of the sea has become a major concern. While most landlocked bodies of water are embedded within one or maybe two countries, the Caspian Sea borders five countries (Azerbaijan, Russia, Kazakhstan, Turkmenistan, and Iran). Three of these countries (Azerbaijan, Kazakhstan, and Turkmenistan) only recently received independence from the former Soviet Union, in the 1990s. For this reason, the political relationship between these countries and Russia is a bit sensitive. Until the breakup of the Soviet Union, almost all of the Caspian Sea belonged to Moscow. The dispute over the Caspian Sea has become a forum for these newly free nations to demonstrate their independence, and Russia can only respond with patience.

While ownership has not yet been decided, oil and gas development has progressed. In the late 1990s, discussions about the sea were centered on dividing the body of water into national zones as called for by UNCLOS. The three new countries bordering the Caspian Sea agree with UNCLOS, but this ruling may leave Iran and Russia with barren sections. Naturally, Russia and Iran do not agree with the ruling and have attempted to find a way to make UNCLOS ineffective. So far they have come up with the argument that the Caspian Sea is technically a lake and, thus, not subject to the United Nation's law. The situation has been further complicated by the fact that any oil or natural gas sent from the sea has to travel through one of the bordering countries so an agreement must be made. Regardless of the unsettled nature of the dispute, the three newly independent countries have been granting production rights to oil companies. While outright war has not developed among the countries in the Caspian Sea region, the issue has yet to be resolved.

These issues can be resolved amicably and cooperatively and do not always have to result in one nation's total control of the disputed area. However, the resolution of these disputes depends largely on the extent to which a country desires control of a particular oil-rich region. For a country that is in economic shambles and whose oil fields are in a similar state, withdrawing from conflict is not always easy. In the case of Iraq, leaving Kuwait was difficult and humiliating, but for Saudi Arabia an amicable arrangement was reached. In the 1930s the British established the Saudi-Kuwaiti Neutral Zone due to political conflict. Since the first oil field came onstream in the neutral zone in the 1950s, Saudi Arabia and Kuwait have agreed to share it. The Saudi-Kuwaiti Neutral Zone has flourished, with a major oil field that produces almost 600,000 bbl of oil per day.

ECONOMIC SANCTIONS

Sanctions serve as an often effective tool to place pressure on a political regime deemed regionally or globally dangerous and to draw international attention to a country's violation of human rights. As for directly causing a leader to fall from power or a regime change to occur, sanctions are not effective. Nonetheless, the use of economic sanctions provides the best example of how oil and politics mix. The decision to factor oil prominently into these plans also reinforces our understanding of just how important oil is to a nation's political and economic standing. Often *sanctions* take the form of an internationally accepted commitment, usually implemented by the United Nations, to place pressure on a country without using direct armed force. These sanctions usually involve the restriction of international trade and finance. Sanctions include targeted measures such as arms embargos, travel bans, and diplomatic restrictions. In regard to oil, a country generally faces an embargo on the export and import of petroleum products. Oil embargos have been used against both oil-importing and oil-exporting countries. A prominent example of an embargo on an oil-exporting country is the case of Iraq during the 1990s. The most well known example of an oil embargo against oil-importing countries is the 1973 embargo by Arab oil-producing countries against the United States and select European supporters of Israel (for more on this embargo, see Chapter 4). Aside from the 1973 embargo, most international sanctions are implemented by the United Nations.

In the history of the organization, the United Nations has executed fourteen sanctions, with the majority taking place in the 1990s. The most prominent of the sanctions was against Iraq because it was the longest and most economically severe. A few hours after Iraq's invasion of Kuwait in August 1990, the United Nations Security Council (UNSC) condemned the invasion and called for Iraq's immediate removal. When Iraq refused, the UNSC imposed economic sanctions on Iraq and established protected zones ("no fly zones") in northern Iraq to protect Iraqi Kurds and Shi'ite Muslims in the south. Four days after the invasion, the UNSC passed Resolution 665, turning the economic sanctions into an outright blockade, during which all shipments would be closely monitored. Most important, this cut off oil exports leaving Iraq. After Iraq withdrew from Kuwait in February 1991, the United Nations continued to enforce economic sanctions in order to pressure the Iraqi government to improve the treatment of its own people as well as its regional neighbors. As long as Iraq abused the Iraqi Kurds and Shi'ite Muslims, it would face UN sanctions. The hope was that economic pressure would drive the Iraqi people to remove Saddam Hussein from power and install a more reasonable leader. Indeed, this never happened. In the standoff between the United Nations and the Iraqi government, the former compromised its plan of action first.

In 1995, the UNSC adopted Resolution 986, allowing limited Iraqi oil exports that could be used only for humanitarian purposes such as food and other supplies. This resolution, better known as the oil-for-food program, lasted from 1995 until 2003. In 1998, Iraq declared that it would cease all cooperation with the UN arms inspectors and monitors unless the UN embargo was lifted. The United Nations responded by lifting the ceiling on the amount of oil Iraq could export. Close monitoring of trade, however, still continued. Despite the sanctions, it was reported that Iraq smuggled large volumes of crude oil out of the country through its neighbors by truck. Even with the destruction of the pipelines, terminals, and the Khor al-Amaya—Iraq's main export outlet on the Persian Gulf—oil was still smuggled out during the UN sanctions. The profits from smuggling provided Saddam Hussein with billions of dollars. Hence, he remained in control until his forced removal in 2003. The success of the embargo in toppling the Iraqi regime is questionable and contributes to the larger debate over how useful economic sanctions involving oil are as political weapons.

Scholars have argued that oil sanctions are not effective because they are inherently problematic and are difficult to sustain. The effectiveness of a sanction begins with the commitment of the global community to the political cause. For this reason, the United Nations does not act on every oppressive regime that comes into power in the world. There has to be level of economic or emotional interest on behalf of the countries imposing the sanction. For example, the United Nations rallied support for sanctions against South Africa with ease. Members used sanctions against South Africa in order to place pressure on South Africa's apartheid government, and Nigeria became one of the campaign's most outspoken members. The strength of sanctions really comes from the emotional connectedness that a country feels toward the situation.

Between 1948 and 1994, South Africa was ruled by a white minority that considered itself superior to the black majority. The term *apartheid* is an Afrikaans word meaning "separation" or "apartness," but within the context of South Africa came to mean a legalized system of segregation. Apartheid is said to have ended when Nelson Mandela was released from prison and sworn in as president in 1994. The United Nations found the problem of South Africa a long and difficult battle. In 1962 the United Nations passed Resolution 1761, which publicly condemned South Africa's racist apartheid policies. To add strength behind the policy, the United Nations called on its members to cease military and economic relations with South Africa. The sanctions included the prohibiting of imports, exports, and bank loans. Under the broad prohibition of imports were petroleum products, which oil-producing countries such as Nigeria took seriously. In 1973, the United Nations stepped up the pressure by drafting the International Convention on the Suppression and Punishment of the Crime of Apartheid, which created a legal framework for applying sanctions. Among the oil-producing countries that took a particularly active role in this campaign was Nigeria.

The situation in South Africa represented an outstanding colonial presence within Africa. At the time, Nigeria acted as one of the most active proponents of an independent, black Africa where the legacy of white domination would cease to exist. For this reason, Nigeria stood at the front line as one of the few developing African countries. In fact, Nigeria was one of the few African countries to host two UN conferences regarding the situation in South Africa. Like Nigeria, many other countries in the world felt quite impassioned by the situation in South Africa. The United States, having struggled with racial oppression and legal segregation for decades, took an active role in eliminating racism all over the world. It has been argued that the United Nations as a whole acted relatively swiftly to the situation in South Africa, while more slowly toward hostile situations elsewhere. In the case of Iraq, for example, Saddam Hussein's brutal reign began in 1979, but the United Nations did not respond to the atrocities until the invasion of Kuwait in 1990. The United Nations' selection process has raised questions over what regions and portions of the global population are allowed to suffer.

Economic sanctions are also considered ineffective because it is almost impossible to effectively execute an embargo or sanction on a specific country using oil because oil is a truly global industry. For example, crude oil may come from Nigeria, but is refined in Europe, traded on the NYMEX floor, and consumed in California. By the time it reaches the consumer, it is difficult to determine its origin or what company produced it. Despite the widespread effort, the UN-imposed sanctions did not bring down the apartheid regime in South Africa. With a large supply of coal available, South Africa did not depend heavily on oil imports at the time. Also, independent producers delighted in selling oil at premium prices to South Africa. In fact, Nigeria's decision to nationalize British Petroleum's holdings in 1979 partially stemmed from the oil company's practice of taking the Nigerian oil it produced and selling it in South Africa. As we saw in the case of Iraq, the sanctioned country smuggled crude oil out and sold it. An oil embargo on an oil-importing country only encourages it to buy from countries that ignore the UN-imposed sanction. The end result is that the sanctions do not greatly curb the outgoing oil or the incoming profits, which often go directly to a brutal leader who, in turn, invests in greater military strength.

Oil sanctions are also not effective because they usually affect the wrong group of people. In the case of Iraq, the standard of living dramatically declined, while the population increased. To manage the small amount of food coming into the country, Iraq adopted a rationing system. The portion that people received was only a small portion of what the human body needs to survive. As a consequence, the country was near a state of famine. Even during the oil-for-food campaign, the much-needed supplies did not always reach the Iraqi people. By the mid-1990s, it became apparent that the economic sanctions hurt the innocent Iraqi population instead of the government. Fortunately, the removal of Saddam Hussein also meant the removal of the UN sanctions on Iraq, which had been in place for more than ten years. In May 2003, the UNSC lifted all trade and financial sanctions imposed on Iraq. It also called for the creation of the Development Fund for Iraq, which took proceeds from Iraqi oil sales and invested them in a new post–Saddam Hussein Iraq.

Serious debate has arisen within the past ten years over the true effectiveness of sanctions pertaining to oil. In most cases they do not bring down the ruthless political heads at which they are targeted. In reality, sanctions magnify a regional problem into an international one. When the United Nations asserts its sanctions, countries adjust their trade and finances to accommodate the policies. Even countries that ignore the sanctions are affected by them. In essence, oil-related conflict cannot help but spill over into uninvolved countries.

SPILLING OVER

The conflict in Iraq has affected the whole world. It has changed the political dynamic within the Middle East, caused the price of oil to fluctuate dramatically, and disrupted the global supply of oil. Oil producers not involved in the conflict such as Venezuela and Mexico began increasing production to make up for the loss from Iraq. Likewise, something so localized and seemingly innocuous as the construction of a pipeline in the Caspian Sea region can change the relationship among neighboring countries for better or worse and affect the supply of oil to the world's consumers. Today there is hardly a corner of the world where oil has little or no impact. Indeed, oil and politics are intimately connected.

To the public, the role of oil as a source of conflict features much more prominently in global news than its role as a basis on which alliances are built. This is because oil-related conflicts seem to have a severe and extensive impact. These conflicts vary from territorial disputes to domestic clashes to outright war on an international scale. In each case, spillover often takes place in the form of economic, political, and physical intrusion. For example, when Iraq invaded Kuwait, Saudi Arabia became worried about the safety of its oil fields within Kuwait and within its own borders. Because of the Saudi-Kuwaiti Neutral Zone, Saudi Arabia had a vested interest in what happened to Kuwait. Saudi Arabia ran the risk of not only losing oil revenues, but also experiencing an invasion from Iraq as well. In the Middle East, any physical damage caused by war is a severe blow to a country's economic stability. These countries are heavily dependent on oil revenue and must make protecting the oil industry their highest priority.

Increasingly, however, countries engaging in oil-related alliances or conflict have a new issue to consider—the environmental impact of their actions.

FURTHER READING

For more on oil conflict, see Michael T. Klare's *Resource Wars: The New Landscape of Global Conflict* (New York: Metropolitan Books, 2001). For more information on the San José Accord, see Markos Mamalakis, "The New International Economic Order: Centerpiece Venezuela," *Journal of Interamerican Studies and World Affairs* 20, no. 3 (August 1978): 265–95. Several works have been written on Iraq, including Abbas Alnasrawi's *Iraq's Burdens: Oil, Sanctions, and Underdevelopment* (Westport, CT: Greenwood Press, 2002) and Anthony Arnove's *Iraq Under Siege: The Deadly Impact of Sanctions and War* (Cambridge, MA: South End Press, 2002). For more on the Biafra War in Nigeria, see Toyin Falola's *History of Nigeria* (Westport, CT: Greenwood Press, 1998). For more on Chechnya, see Anatol Lieven's *Chechnya: Tombstone of Russian Power* (New Haven, CT: Yale University Press, 1998), and Dale R. Herspring's *Putin's Russia: Past Imperfect, Future Uncertain* (Lanham, MD: Rowman and Littlefield, 2003). For an in-depth look at the relationship between oil and apartheid in South Africa, see Arthur Jay Klinghoffer, *Oiling the Wheels of Apartheid: Exposing South Africa's Secret Oil Trade* (Boulder, CO: L. Rienner, 1989). See also Lindsay Michael Eades, *The End of Apartheid South Africa* (Westport, CT: Greenwood, 1999). On Mexico's role in NAFTA, see Peter H. Smith's "The Political Impact of Free Trade on Mexico," *Journal of Interamerican Studies and World Affairs* 34, no. 1 (Spring 1992): 1–25, and M. Delal Baer and Sidney Weintraub's *The NAFTA Debate: Grappling with Unconventional Trade* (Boulder, CO: Lynne Rienner, 1994).

Chapter 6

Environmental Concerns

It wasn't the Exxon Valdez *captain's driving that caused the Alaskan oil spill. It was yours.*

Greenpeace, 1990[1]

As much as oil-producing countries appreciate the prestige and the revenue that their oil industries generate, they also realize that the industry has a strong negative impact on their national ecology. Air, water, and land throughout the world have been the victims of pollution and destruction. The environmental damage caused by oil-related accidents or deliberate destruction of oil facilities can no longer be ignored. Every year governments and the international community are becoming less tolerant of oil spills, the clearing of valuable tropical forests, and the displacement of communities because of water, air, and soil contamination. International organizations and oil-producing governments are looking for ways to produce oil in a more environmentally conscious way. In this chapter we will examine the impact oil production, transport, and usage have on the environment as well as governmental and international attempts to lessen the environmental impact of the oil industry through laws and pacts.

Environmental damage from the world oil industry is a cumulative problem that starts with the world's demand for this nonrenewable resource. As the human population grows, the demand for petroleum-based goods increases. While the growth rate of the world's population in the past year has slowed, it is still expanding. In the past fifty years, the population has nearly tripled, making today's total population just over 6 billion. The population factor, of course, explains only part of the growing demand for petroleum products. The other factor is the spread of industrialization and wealth, which have increased the demand for petroleum products. More and more families own multiple vehicles, have an improved standard of living, and are using more petroleum-based products than before. Indeed, this second factor is just as difficult to control as the first.[2] The oil industry, no matter what the level of precaution may be, is an environmentally destructive business. While it seems a difficult task to reduce the global population and our reliance on petroleum, we can at least try to minimize the impact of oil production and usage.

The environment has an incredible ability to rejuvenate and cleanse itself from many destructive actions (natural or human-made). While one small oil spill may not alter the landscape and wipe out entire species of animals permanently, several oil spills, blowouts, and pollutant emissions into the air within one region or body of water can. Many oil refineries and production wells have been in constant operation for more than fifty years, contributing to environmental decline a little more every day. But what damage, exactly, does the global oil industry cause? While some generalizations can be made, the impact on a specific area depends on the location and the operations being conducted there. The impact largely depends on the fragility of the environment. Venezuela's tropical forests, the home to thousands of rare plants and animals, need more protection than Saudi Arabia's Empty Quarter, which is home to virtually no animal or plant life. The world's water and air, however, travel without regard for national boundaries, making the overall impact on these resources more uniform.

Assessing the environmental destruction that occurs can be difficult because the information has to come from a country willing to either do the research itself or allow an international agency into the country to do it. Few countries want to report their emissions and oil leaks, especially when they are high; nor do they want an exposé written about the damage they inflict and have it published worldwide. Atrocities reach international ears when an underground environmental group brings the world's attention to the complaints of a local community. The central government's goal usually is to contain the situation. In short, oil-producing countries have little incentive to quantify the damage. For this reason, specific information varies among oil-producing countries. A great deal of the damage occurring in Nigeria is known because community groups took their grievances to the world and environmental activists all over the world joined their resistance movement. In Iraq and Saudi Arabia, however, the extent of the damage is less clear. Partially this is because much of the Middle East consists of uninhabited desert and the direct impact on people is minimal. For these reasons, assessing the environmental impact of an oil industry can be difficult, but it is by no means impossible.

IMPACT ON THE ENVIRONMENT

The global oil industry's impact on the environment can be divided into two parts: immediate and long-term effects. The immediate environmental consequences of the global oil industry include the day-to-day impacts, such as water pollution, land use, land clearing destruction, industrial waste disposal, and air pollutants. These impacts include those that are noticed by the local community and affect the population's everyday lives.

During prospecting, seismic studies require underground explosions to locate oil reserves. Getting to these sites often requires the clearing of forests and the construction of trails and roads, as well as the setting up of camps equipped with comfortable living amenities. The amount of land transformation only increases once oil is discovered and production begins. Pipelines and the necessary pumping stations require clearing and construction as well. Roads connecting remote fields to refineries need to be built, as do offices and storage tanks. Once these facilities are up and running, there is the risk of accidents, leaks, and spills, which are too often everyday occurrences. In addition to daily small instances of damage are catastrophic ones, which always make the headlines.

The catastrophic impacts are very dramatic and usually result in the loss of animal and plant life. These usually include onshore and offshore *blowouts*, which are explosions and fires at refineries, oil rigs, storage tanks, and pipelines.[3] A notable example occurred in Mexico in 1978 in the Bay of Campeche. An oil well malfunctioned, and during attempts to fix it and restore oil circulation to the system, oil began to flow out of control, resulting in a blowout that lasted nine months and twenty-two days. The blowout dumped more than 3,000 bbl of crude oil into the bay and destroyed part of the platform (what was left of the platform was intentionally sunk).[4] A second type of catastrophic impact is oil tanker accidents during oceanic transport. For example, Exxon's tanker, *Valdez*, collided with a reef near Prince William Sound, Alaska, in March 1989. Contact with the reef created a hole in the tanker, causing a spill of almost 300,000 bbl of crude oil. Similarly, an independently-owned tanker, the *Prestige*, split in half, spilling oil off the coast of Spain in November 2002. Investigations concluded that the twenty-six-year-old tanker's age and declining condition contributed to its failure. Since its sinking, thousands of barrels have continued to wash up on shore in Spain. Incidents such as these have raised public awareness as well as corporate and government awareness, since cleanup projects cost millions of dollars. A third catastrophic impact is that of deliberate destruction.

Deliberate destruction varies from tapping into oil pipelines to siphon oil for private sale, causing incidental destruction, to deliberate sabotage of oil facilities. News reports come out all too frequently about civilians siphoning oil from a pipeline and causing a major blowout. Siphoning primarily occurs in poverty-stricken oil-producing countries; civilians take oil to sell privately on often flourishing black markets. Siphoning is not only illegal, but also very dangerous. One spark from attempting to create a hole in a pipeline can create huge flames that are difficult to smother. Also, the infrastructure to respond quickly and effectively to such a disaster often does not exist. This has become a particularly severe problem for Nigeria, resulting in several disastrous pipeline explosions. In 1998, an explosion near the town of Jesse within the Niger Delta took the lives of more than 500 people. People had been collecting leaking oil from the pipeline for weeks until a spark from a motorbike ignited it.

In addition to siphoning, people respond to economic and political instability by damaging pipelines. This second situation occurs as a civilian protest against the government or the foreign oil companies operating in the community. Oil pipelines are usually the main target because in many cases they carry oil and natural gas through remote parts of the country where little to no security exists. Pipelines can be hundreds of miles long and are often aboveground to allow the steel to expand and contract depending on the weather. In countries such as Saudi Arabia and Iraq, these pipelines run across large expanses of uninhabited desert. By damaging a pipeline, the saboteurs can halt the national industry and call international attention to their plight. Pipeline sabotage acts like an economic sanction in that it places severe economic pressure on the government. By damaging a pipeline, people literally cut off the government's economic lifeline. In 2004, Iraq saw repeated attacks on pipelines, which nearly ground the industry to a halt. The Iraqi interim prime minister, Iyad Allawi, reported a loss of more than $200 million in revenues. During that year, Iraq's industry crawled along at around 1 million bpd compared to its prewar 3 million. For every example of pipeline sabotage, the country suffers a slowdown in production, massive environmental destruction, and the deaths of many innocent people.

The dumping of oil and the bombing of fields and facilities have also become weapons of war. Iraq adopted these practices as part of its military strategy on two occasions. The environmental destruction, in both Gulf wars, began with heavy military vehicle traffic over delicate desert landscape and included the scars left behind by explosions. During the Iran-Iraq War (1980–1988), Iraq intentionally torched three Iranian oil and gas wells in 1987. A few years later, Iraq utilized this same method of warfare against Kuwait in the Gulf War (1990–1991), beginning in January 1991. As many as 11 million bbl of oil were dumped into the Arabian Gulf. Miles of beach were covered in oil, severely damaging Kuwait's shrimp stocks and seabird population. Also, several oil fields (more than 600 wells) in Iraq and Kuwait were torched. These fires, known as the Kuwait oil fires, burned for months, with the last of them extinguished in November 1991. Within a few months, more than 140,000 bbl of crude oil polluted the atmosphere, falling back down in the form of soot not only Kuwait but also on its neighbors.[5] While this had a traumatic effect at the time, what are the long-term effects of such catastrophic disasters and day-to-day pollution?

Land Pollution

Land pollution poses a problem for local communities that depend on the soil for farming and herding of livestock. It also harms vital forms of plant and animal life living within the ecosystem. In many oil-rich countries, people continue to live in traditional ways: they depend on the land for shelter, food, and medicine. For the soil, the most damaging effect of the oil industry comes from the refineries. The solid and liquid wastes from the refining process are sometimes released into the ground through poorly constructed reservoirs, which are meant to contain the waste.

Crude oil leaves a sticky residue, which can harm animals. Thick oil can stick to an animal's feathers or fur, which can increase its body temperature, causing it to suffer from overheating. The development of an oil industry encourages deforestation, particularly in tropical regions. In Venezuela, for example, deforestation is occurring at a higher rate than in the rest of South America because of the expansion of the oil industry within the Orinoco Belt. Oil workers clear the forest in order to make the transportation of equipment easier. Unfortunately, oil-producing countries are often also rich in rare natural beauty. The latest site of interest for oil production has been in south central Russia, near Lake Baikal and Tunka National Park. This region includes one of the largest wetlands in the world and contains ancient forests and endangered species. Lake Baikal, commonly referred to as the "Galapagos of Russia," is home to rare plants and animals, including the freshwater seal called the nerpa. The lake's ecosystem risks environmental pollution from one of Russia's largest oil fields. Tunka National Park includes 3 million acres of unspoiled forest and is home to a variety of animals such as the snow leopard. Unfortunately, Russian oil companies have plans for running a pipeline through the national park.[6]

Air Pollution

Air pollution is caused by the emission of pollutants from refineries, gas flaring at production wells and refineries, and the use of petroleum as fuel. Among other pollutants released into the air, a great deal of concern exists over the emission of greenhouse gases (i.e., carbon monoxide, carbon dioxide, methane, nitrous oxide, and several others). These gases are believed to be accumulating in the atmosphere and upsetting the earth's temperature, causing *global warming*. This is by far the most perplexing and controversial international environmental topic because it is the least understood by scientists and the general public. Many believe that the carbon dioxide, we emit through everyday human activity—transportation, manufacturing, agricultural activities—goes into the atmosphere and traps heat, thus creating a *greenhouse effect*. The term is derived from how a garden greenhouse functions (the glass enclosure lets in light to warm the plants and the soil, and traps the heat). As a result of this greenhouse effect, the annual average surface temperature of the Earth since the Industrial Revolution (late nineteenth century) has increased. It is believed that the rate at which global warming occurs is directly related to the amount of carbon dioxide released into the atmosphere. In other words, the greater the consumption of petroleum products, the warmer the Earth's surface becomes. Warmth causes the world's glaciers to melt, sea levels to rise, and a change in precipitation patterns and overall climate.

Water Pollution

Damage to global water sources occurs during several stages of the oil production process. Overall, it is estimated that about 95,000 bbl of oil are spilled annually into the world's rivers, lakes, seas, and oceans.[7] Recent studies have indicated that more than half of this pollution comes from cleaning and discharging operations on oil tankers. When oil fields mature, and their rate of output decreases, oil companies often have to inject water into them to help force out more oil. In these operations, freshwater resources may be overused. Day-to-day activities such as these may seem minor, but the cumulative effect is significant. Most of this damage takes place during the refining process, when oil waste seeps into the ground and settles in underground aquifers and drinking wells. Leaks from oil pipelines also contribute to water pollution. One example of indirect destruction and consequent water contamination happened in Russia in 1994 when a leaking pipeline near the town of Usinsk in northern Russia (Komi Republic) was temporarily contained by a dike— until the barrier collapsed. The Russian oil company Komineft operated the pipeline, which runs between the Kharyaga oil field and the city of Usinsk. The dike collapsed because of snow and subartic temperatures and poured around 750,000 bbl of oil on the Siberian peninsula. Because of the frozen ground, the oil did not soak into the ground, but instead traveled long distances, flowing into the Kolva River and eventually into the Barents Sea. Accidents such as this have long-lasting effects on the world's bodies of water.

The catastrophic oil spills during transportation, however, receive the greatest media attention. In addition to tanker spills by the *Valdez* and *Prestige* mentioned earlier, another similar one occurred near Angola. In 1992 the independently-owned *ABT Summer* spilled more than 35,000 bbl off the coast of Angola, in southern Africa. In addition to tanker-related pollution, production activities on- and offshore also contaminate waterways. Oil leaks and spills such as these kill seaweed and valuable microscopic plants and animals that sea creatures consume.

The length of time that oil pollution impacts land and water is a subject of some debate. Many researchers emphasize that over time, crude oil in water undergoes biological and chemical processes that change its composition and environmental impact. On land, the spilled oil eventually evaporates or dissolves, and disperses. The overall idea is that what is spilled will return to the subsoil from which it came. But how long does the rejuvenation process take? Some argue that although this may, in fact, be the case, no knows how long the process takes or what the long-term results may be. Also, the rejuvenation process does not lessen the importance of taking greater preventive steps toward pollution. As evident in studying the various forms of pollution caused by the global oil industry, the impact is quantifiable—barrels spilled, amounts of animal and plant life destroyed, amount of carbon dioxide emitted into the air. Neither the source nor the impact of environmental damage is a mystery. Why, then, are governments of oil-rich countries so slow to protect the environment?

SLOW TO GO GREEN

The twentieth century represented the beginning of an understanding that industry had detrimental effects on natural surroundings. People began to see that industrial smoke and waste affected the water they drank and the air they breathed. At first, it was believed to be a problem that affected only the urban and industrial areas of a country. But when large-scale factories and oil production moved into rural communities, these areas became affected as well. In fact, we now know that the the most threatened areas are forested regions inhabited by small communities. The dumping of industrial waste into the rivers and lakes threatens water supplies and aquatic life. With the rise and expansion of industrial activity throughout the world, governments began to monitor and limit the amount of pollutants industries could release. Unfortunately, oil-producing countries' commitment to the environment varies and, in most cases, exists only on paper. Nonetheless, the public recognizes that problems such as global warming can be addressed only at a global level.

Leaders of oil-producing countries often ignore warnings of environmental damage from international organizations (governmental and nongovernmental groups alike). There are two reasons for this response—or lack thereof. First, much of the criticism comes from the West, countries of which, in many cases, acted as former colonial powers for many developing oil-rich countries. Iraq and Nigeria, for example, fought hard to remove British control. While the United States never officially claimed any colonies for itself, several oil-producing countries feel as though the U.S. corporate presence differs little from colonialism.

Second, many oil-producing countries view environmental concerns as a wealthy country's luxury. Countries struggling to develop often resent what they perceive as the West's overconcern for how they use or abuse their environment. Developing countries argue that the West caused severe environmental damage when it developed and industrialized in the early twentieth century with air and water pollution as well as deforestation. These countries resent the concern of the West for the preservation of their land. They believe that environmental restrictions on their industries are not helping them to grow economically. In fact, some believe that environmental standards are a way of preventing developing countries from growing economically.

Third, most developing countries are more concerned about food supply and employment than about the harmful effects of their oil industries on the environment. The need for economic growth and societal demands often conflict with environmental concerns. Many governments depend heavily on oil revenues and are reluctant to divert some of the money into environmental safeguards. Aside from eating up revenue, high environmental costs may deter investors. The unfortunate truth is that many of these countries attract foreign investors because of their relaxed environmental codes. The West, in effect, appears to be standing in their way. For this reason, many of the developing countries have not signed international treaties or drafted any sort of serious plans to reduce carbon dioxide emissions or water pollution. In December 2003, the former secretary-general of OPEC, Dr. Alvaro Silva-Calderón, clearly expressed this sentiment to the UN Framework Convention on Climate Change:

We are unhappy about calls for new commitments to be made by developing countries, which would affect the ability of many sovereign states to achieve sustained economic growth, develop their social infrastructures and eradicate poverty. We insist once again that oil-producing developing countries do not end-up as net losers from the climate change negotiations. We are still not satisfied that our legitimate concerns about the adverse impact of response measures on our hydrocarbon-dependent economies have been properly addressed, in spite of the provisions made for this in both the Framework Convention and the Kyoto Protocol.[8]

Finally, oil-producing countries assert that the root of the problem does not actually lie with them. In many oil-producing countries, who actually produces the oil, causes the spills, and uses unsafe methods of disposing of industrial waste? The oil companies themselves do. Overall, the oil companies have come under severe criticism for not taking a leading role as the proponents of safe oil operations. This is because they not only control much of the world's oil production, but also because they have the expertise, research facilities, and money to do so. Pressure groups more recently have targeted their demands for change toward the oil companies. As a result of a growing body of laws, international oil companies may face liability for matters such as oil spills and may be forced to pay the cleanup costs. In 2004 environmental groups began gathering evidence to place partial blame on ExxonMobil for global warming. This marks what could be a legal milestone for environmentalists. The groups are accusing the company of causing a significant increase in carbon dioxide emission—from customers who used its oil—in the past 120 years.[9] In other words, the efforts exhibited by the oil companies thus far have not been considered adequate by everyone.

Oil-rich countries also view the root of the problem as lying in the West because the West overwhelmingly consumes the most petroleum products in the world. Oil-producing countries argue that they are forced to produce more every year to keep up with demand. According to the Energy Information Administration (EIA), world oil consumption rose in 2004 by about 1.9 million bpd. Out of the world's consumption, industrialized countries such as the United States collectively account for 55 percent of this increase. Demand among developing nations increased by 0.7 million bpd. Latin America and the former Soviet Union increased only slightly from previous years. Overall, predictions have been made that by the year 2025, world consumption will reach nearly 120 million bpd.[10] With these kinds of estimates, environmental activists feel that global intervention through laws and agreements is essential.

INTERNATIONAL PRESSURE GROUPS

While the United Nations represents the only international lawmaking body, pressure groups have raised a great deal of awareness through protest and research. The influence of pressure groups on the United Nations cannot be denied. Because of these groups, severe environmental destruction in often remote corners of the world has garnered international attention. Known for their grassroots, ground-level approach, these groups often have a better sense of the everyday destruction than do national governments or the United Nations. In many ways they provide an international voice for minority ethnic groups who lack access to media tools or international connections. Because these groups operate independently, they move freely in focusing their campaigns and protests. Many groups emerge out of a need to draw international attention to oil-related destruction in their communities, while others focus on every aspect of environmental destruction, including oil. Pressure groups often take a multipronged approach to getting their message across by placing pressure on oil companies and national governments, and by appealing to international institutions for a legal response. Hundreds of organizations exist that operate on local, national, and international levels. Here are just a few that through their work have gained international notoriety.

Oilwatch International

Oilwatch International is an organization that dedicates its efforts to ending the activities of the destructive activities of the oil companies in tropical countries. The organization formed in 1996 and is based in Quito, Ecuador. The group's objectives include creating a plan of action against oil companies who are disrupting, or could potentially disrupt, the ecology within a community. Its members include activists from Brazil, Cameroon, Colombia, Gabon, Guatemala, Mexico, Nigeria, Peru, South

Africa, Sri Lanka, and Thailand. It maintains a network for concerned activists interested in finding up-to-date information on oil company operations and the resistance movements to them within tropical countries.

Greenpeace International

Greenpeace International is the most well known environmental rights organization because it campaigns against all forms of environmental destruction and abuse, not just those pertaining to oil. It is famous for its attention-grabbing campaigns throughout the world. The organization began in 1971 and has chapters all over the world. Its main objectives include promoting the use of renewable sources of energy and protecting delicate ecosystems. In relation to oil, Greenpeace played a crucial role in the implementation of a fifty-year moratorium on mineral production in Antarctica.

Project Underground

Project Underground began in 1966 to expose the environmental abuse and human rights violations that exist within the global oil industry. The main objective of the organization is to provide legal information to communities threatened by destructive extractive industries. One of their primary concerns has been alerting these communities to their legal rights under international law.

Friends of the Earth International

Friends of the Earth International (FOEI) is a worldwide federation of national environmental organizations. The organization formed in 1971 by four organizations representing France, Sweden, England, and the United States. Today the organization includes roughly sixty-eight groups and is based in Amsterdam. The focus of the organization is to protect the earth, to repair damage inflicted on the environment, and to promote eco-friendly sustainable development on a global level. The member groups are united by a common commitment to grassroots activism, which includes persistant compaigning by a group as a member of FOEI and as an independent organization.

Environmental Rights Action

The Environmental Rights Action (ERA) serves as the main activist group of Nigeria. In 1993 the organization formed to campaign against the environmental and social crises that occurred in the Niger Delta (see Chapter 7). For this reason, the organization is based in the city of Port Harcourt, which lies in the center of Nigeria's oil-producing region. In addition to advocating environmental rights in Nigeria, the ERA is also involved with Friends of the Earth International and Oilwatch International. This remarkable organization was the recipient of the Sophie Price award in 1998 for its dedication to defending the rights of people and their environment.

Organizations such as those mentioned above often represent the spark of national and international response to environmental issues. Through lobbying, protests, and international exposure, these groups often use an exposé approach to place the governments, companies, and international organizations in an uncomfortable position. None of these institutions desires bad publicity for polluting drinking water and destroying ancient forests. Pressure groups recognize this and use it to prod them into action.

NATIONAL RESPONSE TO PROTECTING THE ENVIRONMENT

Oil-rich countries have responded to pressures to improve their activities and the activities of the oil companies within their countries through the use of modern technology and laws. The legal response to pollution problems differs among countries, regions, and international organizations. Many developing countries lag behind accepted policy and practice in the industrial world by about twenty years because of a shortage of expertise, governmental infrastructure, and, perhaps, interest.

Some developing oil-producing countries have made concerted efforts to improve the air quality and lessen the environmental destruction through passing regulatory measures such as environmental quality, emission, and technology standards. Many countries also use fuel quality control, which includes limiting the lead levels in gasoline. Restricting when and where cars can be driven also serves as an incentive for people to change their preferred means of transport. For example, Mexico has attempted to curb its fossil fuel emissions by limiting the number of cars operating (based on license plate number) on a given day in Mexico City, its most heavily populated and polluted region. Transportation emissions account for 70 percent of local air pollution in Mexico City. In Venezuela, the government launched plans to reduce levels of lead in gasoline and control emissions from public transportation vehicles. National oil companies have begun to take responsibility for environmental destruction caused by their oil operations. In 2004, Mexico's environmental protection agency, Profepa, received over $7 million to repair the environment from oil-related damage. Funded by PEMEX, the plan included the planting of thousands of trees and establishing new levels of environmental protection.[11] Most developing countries, including Russia, recognize natural gas as a valuable and cleaner source of energy.

The importance of natural gas has changed over time. Oil-producing countries within the past ten to twenty years have watched their major oil fields mature and their production levels drop as no new crude oil fields are found. What they have realized recently is the potential of natural gas, which is found in abundance within crude oil fields as well as in fields of natural gas alone. Until recently, oil companies burned off the natural gas during oil production without much thought of capturing and using it as a source of energy. Over the years they developed the technology to reinject the natural gas into mature gas reservoirs as a way to enhance oil recovery and to use it as a form of fuel. Environmentalists encourage the use of natural gas because it is a cleaner alternative to petroleum products. It is desirable because it is easy to combust since it mixes readily with air, and it produces three-quarters the amount of carbon dioxide of oil. In some countries, cars have been fitted to run on natural gas by adding a pressurized tank for compressed natural gas storage and a fuel line for the gaseous fuel. Natural gas vehicles have lower fuel costs and have the advantage of contributing to improved air quality. Countries such as Mexico are adopting this form of automobile fuel. Picking up on the rising trend of environmental consciousness, oil-related companies have tried to offer new eco-friendly products.

CORPORATE RESPONSE TO PROTECTING THE ENVIRONMENT

The life of any company involves knowing and responding to what consumers want from a product. Companies spend billions a year on marketing research to identify the latest trends. What oil companies and car companies found is that in the 1990s, the average consumer had become more environmentally conscious. Vehicle companies responded to this by developing the technology to reduce carbon dioxide emissions and rely less on petroleum products for fuel. Within recent years, electric cars and hybrid cars (a combination of a rechargeable battery and fuel) have been available on the market from Toyota and Honda. Amid the sport utility vehicle (SUV) craze, particularly in the United States, Ford and Lexus have drafted plans for a hybrid model to be available in late 2004. In fact, Ford launched its model in 2005 making it the world's first hybrid SUV. By developing these new vehicles, these companies may gain a competitive edge. Consumers today have a wide variety of choices for almost any product in the world. This reality does not escape the oil companies.

People all over the world make decisions to buy certain brands of petroleum products based on availability, convenience, quality, and social and political reputation. While the oil companies focused on the first three, they did not always consider the last. In the 1990s, oil companies began to realize that their reputations as socially and environmentally friendly organizations factored into consumers' choices. As a result, the oil companies began to be more conscious of their actions and to make sure this came across in marketing their products. These companies have tried to commit themselves to better pipeline management and environmentally sound extraction and refining processes. They have also developed new ways to not only reduce the amount of industrial waste, but also recycle it. Scientists have received funding to research a special type of microorganism that digests the oil and breaks down hydrocarbons at the molecular level.

While developing countries have taken some initiative in fostering change for the improvement of environmental standards, many governments and organizations agree that it is not a national issue. Many of the pressing environmental concerns, particularly global warming and water pollution, are international problems; they are rarely confined to a single country. Pollutants often migrate, creating problems in neighboring countries as well as in the country in which a spill took place. The question is whether environmental law should become an international affair or remain a domestic one. Although environmental issues are not new to international relations, world leaders increasingly have moved environmental issues from the periphery to the center of their political agendas. They recognize these problems as having truly global dimensions and are increasingly supplementing their laws with international treaties.

THE ROLE OF INTERNATIONAL ORGANIZATIONS AND INTERNATIONAL LAW

International organizations and lawmaking bodies have come to realize the importance of their roles in three distinct ways: first, to place pressure on oil-rich countries (developing and developed alike) to raise their environmental standards; second, to place pressure on oil companies (multinationals and state-owned alike) to operate using more environmentally friendly methods; and third, to upgrade the level of regulation and protection of international waters. Recognizing the need for action, the United Nations and the World Bank established the Global Environmental Facility (GEF) in 1991 as an organization devoted to international environmental issues. The goals of the GEF include addressing issues of biodiversity and climate change, among others. One year later, the GEF launched the largest international conference—the United Nations Conference of Environment and Development (UNCED)—better known as the Earth Summit.

At the Earth Summit, representatives from 179 nations (including heads of state and nongovernmental organizations from around the world) met in Rio de Janeiro, Brazil. Out of this conference came the Rio Declaration on Environment and Development, which was a collection of twenty-seven non–legally binding agreements. The declaration called for the establishment of a new global partnership based on equality and cooperation. By signing the agreement, member countries agreed to implement international agreements to protect the environment and promote development. While the declaration seemed ambitious, it had one crucial problem: several key signatures from oil-producing countries were missing. Nonetheless, those in attendance went away from the Earth Summit having also created the United Nations Framework Convention on Climate Change, which took place simultaneously, but separately from Rio. The treaty established a global commitment to end global warming. After the Earth Summit, several other plans for global action in the name of environmental protection and economic development emerged.

The Convention on Climate Change and the Kyoto Protocol represent major milestones in the history of environmental protection. The 1992 Convention on Climate Change, which served as an international commitment to reduce greenhouse gas emissions by 2000, but since it was only a nonbinding "aim" of the convention, several countries such as the United States and Canada reported that they could not make the commitment. As the 1992 meeting's goals went unfulfilled, another more promising meeting convened. In 1997 the United Nations held another environment-focused meeting, this time in Kyoto, Japan, as part of the series of Conventions on Climate Change. What emerged from this meeting was the framework for the Kyoto Protocol, which, if ratified, will be a legally binding agreement for countries to reduce their greenhouse gas emissions from

2008 to 2012. All countries would be required to reduce their emissions to a collective 5 percent. The difficulty of the protocol has been forming an emissions reduction level that is feasible for each country. For example, the entire European Union is collectively committed to an 8 percent reduction, the United States 7 percent. The Kyoto Protocol also proposed ideas of how to encourage countries and companies to "go green."[12]

Two important components of the Kyoto Protocol were set up to reward countries and companies and, at the same time, allow a bit of negotiating room. Instead of establishing laws and ways to punish those who violate them, the United Nations envisioned a more positive strategy. One strategy is *emissions trading*. Each company or country has a set emissions allowance. If a company, for example, wants to reduce its emissions more than it is required to, it can sell (i.e., trade) its excess allowance to another company. Countries can do the same. Currently, emissions trading has been implemented in developed countries such as the United Kingdom. A full-fledged international market is the ultimate goal but has yet to reach fruition. For now, emissions trading operates informally and on a small scale. Another strategy proposed by the Kyoto Protocol includes distributing emissions targets with an incentive for a country or company to invest in a project such as planting trees to offset its emissions. This, too, has not gone into full effect.

Many developing countries resisted making a formal commitment and, therefore, are expected to act on a voluntary basis without a specified timetable or target level. For the Kyoto Protocol to go into effect, fifty-five countries must ratify it. Furthermore, countries that emit 55 percent of the world's carbon dioxide (i.e., the United States, Canada, Australia, Russia, and the countries of the European Union and central Europe) must ratify it. The United States withdrew its commitment in 2002, Australia did not signed the agreement, and Russia remained undecided until September 2004. If Russia had not signed the agreement by 2005, the protocol would have collapsed. In February 2005, the Kyoto Protocol went into effect with 141 signees. As of publication India and China did not sign the agreement as well.

Many of the ideas and goals raised at the United Nations' conferences related to the emission of greenhouse gases have reached fruition, but more progress has been made with water pollution. This is largely because water pollution in the form of oil spills is easier to track, quantify, and clean, and, it is therefore easier to press legal responsibility on the party at fault. Up to now, international efforts to protect global seas have focused more on large oil spills than on small, day-to-day contamination because the latter is more difficult to quantify. The overall goal thus far is to make the company that spills oil into water accountable. This, however, is not an easy task. Major oil companies do not always use their own oil tankers to transport oil overseas.

Generally, oil companies hire independent shipping companies to transport their oil and take out insurance to cover their operations in case of accidents, just like any other business. One of the major problems is that oil tankers are often registered in developing countries that have fewer regulations and enforcement capabilities. Countries such as Liberia and Panama are two prominent examples with a significant percentage of all registered oil tankers. In 2001, Liberia accounted for 35 percent of the world's oil tankers. The oil spill off the coast of Spain brought to international attention the complicated and often illegal practice of double registering within the shipping industry. The oil tanker that spilled the oil, the *Prestige*, was a Liberian-owned vessel with a Greek captain and Filipino crew. To make matters more complicated, this ship was also registered in the Bahamas. Oil transport companies double register tankers to lower their operating costs and take advantage of less restrictive laws within developing countries. Also, the liability in case of disaster falls on the ship's crew and not on the oil company. Recently, major oil companies and their insurers have begun to inspect tankers and crews before signing contracts. This strategy protects them from liability in case of oil spills.

If an oil tanker spills oil within a country's territory, then that country has the legal right to hold the oil company responsible. Over the years, countries such as the United States and Spain have tightened their laws to protect their waters, particularly since the numerous spills during the 1990s. In 1958 the United Nations formed the Intergovernmental Maritime Consultative Organization to address the legalities surrounding oil spills. Today this organization, now known as the International Maritime Organization, has adopted uniform international rules and procedures for determining liability for oil tanker spills that include naming the owner of the tanker as the liable party. The company, then, is responsible for bearing the cost of compensation, cleanup, and loss of oil. This is in the case of an oil spill in national territory. In international waters, international law comes into effect. For this purpose, efforts are ongoing to create a uniform law regarding liability that would apply throughout the world. As of 2004, however, this scheme remains only in the development stage.

Oil companies have also stepped up their commitment to reducing environmental destruction deriving from their operations. In 1974 many of the world's largest oil companies formed the International Association of Oil and Gas Producers (OGP). The OGP has fifty-nine members, including Statoil, British Petroleum, Yukos, PEMEX, PDVSA, ExxonMobil, ChevronTexaco, and Saudi Aramco. Collectively, the OGP represents more than half of the world's oil and over a third of its natural gas production. Its members' operations span more than seventy countries. This organization, based in London and Brussels, hopes to conduct oil and natural gas exploration and production in an environmentally respectful way. The OGP acts as a nongovernmental organization with regional and international regulatory groups that work with the United Nations. Within the past few years, responsibility for the environment has fallen not only on the shoulders of the oil companies, but also on the organizations that encourage mining through trade agreements and loans.

Environmentalists argue that the best approach to environmently friendly procedures is for international institutions such as the World Bank Group and the World Trade Organization (WTO) to tighten up their regulations as a way of indirectly enforcing higher environmental standards.[13] Pressure groups argue that until the international community rejects funding for destructive mining projects and ceases to allow goods produced by environmentally hazardous methods to be traded, countries and companies will not alter their harmful operations. The World Bank, activists argue, must reassess what projects it funds. The World Bank did, in fact, entertain arguments that it should cease funding ecologically unsound projects and, instead, focus on renewable sources of energy. Strong resistance from many developing countries has prohibited any development of concrete policies, however. In February 2004, the World Bank reviewed an idea proposed earlier that called for the bank to increase its investment in renewable energy by 20 percent a year. The WTO has faced the same kind of pressure.

The WTO, based in Geneva, Switzerland, is the only international organization that seeks to regulate and oversee global trading activities. The WTO was formed in 1995, taking over the General Agreement on Tariffs and Trade (GATT). It is comprised of 146 member governments (Russia and Saudi Arabia are not members) that reach decisions by consensus. At this time, the WTO has no specific agreement dealing with oil and the environment, but has implemented many provisions on general environmental issues. These provisions include bans or restrictions on certain products. Currently, the WTO's Trade and Environment Committee is more concerned about the violation and enforcement of agreements among countries than about imposing environmently friendly operations on member countries. Environmentalists argue that the WTO is a logical forum for introducing environmental regulations because many environmental concerns are connected to products and services. This is particularly so because these regulations could limit or expand the trade of energy sources. Through these regulations, the organization could encourage the use of renewable energy sources.

RENEWABLE ENERGY SOURCES

For years environmental activists have been encouraging governments and companies to adopt alternative methods of energy, or *renewable energy sources*. These sources include those that provide energy that either can be renewed or are not permanently depletable by production and use, such as solar, wind, ocean, and hydropower. The sun represents the best source of energy because it cannot be depleted by consumption. Overall, renewable resources are more evenly dispersed around the world than oil and natural gas, which only a select number of countries hold. More countries in the world have access directly to large bodies of water, for example, than to oil fields. Equitable allocation of these resources also makes energy more widely available to consumers because of its proximity and resultant lower transportation costs. Renewable sources of energy are used in many parts of the world but are not commonly used in industrialized countries. While petroleum and natural gas are the most widely used sources of energy, neither can be renewed. When an oil or gas field runs dry, no amount of technology can return it to its earlier state. The two main obstacles to the worldwide adoption of renewable energy are the cost and technology to make it feasible. The technology to collect and utilize alternative sources of energy lags and, consequently, makes any form other than fossil fuels an economically and technologically poor choice. Regardless of the present state of the technology, environmentalists are hopeful that alternative sources will replace petroleum in the near future.

FURTHER READING

Several works within the past decade have made enormous contributions to our understanding of the complex relationships among politics, oil, and the environment. For a overview of environmental theory, see Part 2 of Ken Cole's

Economy-Environment-Development-Knowledge (New York: Routledge, 1999). For more information on the environmental impact of the Gulf War, see Muhamad Sadiq and John C. McCain's *The Gulf War Aftermath: An Environmental Tragedy* (Boston: Kluwer Academic, 1993) and Michael T. Klare's *Resource Wars: The New Landscape of Global Conflict* (New York: Metropolitan Books, 2001). For a discussion of the Kyoto Protocol, see Ulrich Bartsch and Benito Muller's *Fossil Fuels in a Changing Climate: Impacts of the Kyoto Protocol and Developing Country Participation* (Oxford: Oxford University Press, 2000). For information on the UN conferences, see Michael Grubb's *The "Earth Summit" Agreements: A Guide and Assessment: An Analysis of the Rio '92 UN Conference on Environment and Development* (London: Earthscan Publications, 1993) and Richard N. Gardner's *Negotiating Survival: Four Priorities After Rio* (New York: Council on Foreign Relations, 1992). For the framework of the WTO, see the WTO's *Understanding the WTO*, 4th ed. (Geneva: WTO, 2003) and its Web site (www.wto.org). For additional information on the Global Environment Facility, see its Web site (www.gefweb.org). For an excellent book linking deforestation and oil, see Sven Wunder's *Oil Wealth and the Fate of the Forest* (London: Routledge, 2003). To read more about the environmental groups mentioned here, visit their Web sites:

Environmental Rights Action (www.essentialaction.org/shell/era/era)

Friends of the Earth International (www.foei.org)

Greenpeace (www.greenpeace.org)

Oilwatch International (www.oilwatch.org.ec)

Project Underground (www.moles.org)

Chapter 7

Human Rights

Lord take my soul, but the struggle continues.
Ken Saro-Wiwa, November 10, 1995[1]

Human rights pertaining to oil has been an issue from the moment the first well was drilled. When oil companies first began producing oil overseas, they had little regard for issues such as the negative impact their operations would have on local communities. As long as they abided by the local government's laws, which were usually few and not enforced anyway, the companies did not concern themselves. During the early twentieth century, virtually no labor laws existed, nor did regulations over the industry's emissions, waste, or handling of the environment. For the oil companies this was an excellent opportunity not only to expand their operations, but also to try new oil extraction techniques and methods of company management without major legal or political constraints.

On the national governments' side, the quest for domestic oil development often overshadowed concern for local communities. Also, national politics and ethnic rivalries played a key role. Often the oil-rich fields existed in remote parts of a country inhabited by minor ethnic groups considered inconsequential and backward by a government ruled by the majority ethnic group. Thus, the governments viewed the arrival of the oil industry as a way of developing new regions. This kind of thinking by the oil companies and the oil-rich countries' governments existed, for the most part, until recently. In some countries, however, minority groups and outside observers believe that this ideology still exists. Furthermore, they see the arrival of oil companies in a less than positive light.

To illustrate this point, we can look at Mexico in the late nineteenth century because it was the first country to which oilmen from the United States went in the late nineteenth century in search of oil. At the time, the Mexican government supported the oil companies and alienated minority groups in the country. In the Mexican state of Veracruz, an ethnic group known as the Haustecs lived as traditional agriculturalists isolated from the rest of developing Mexico. To the Mexican government, the Haustecs, a minority group that occupied valuable land, stood in the way of modernity. When Mexico learned of Veracruz's significant oil reservoirs in 1900, a land grab began. Much of the land was acquired by the oil companies by extralegal means pressuring the Haustec landowners to sell. The oil companies hired agents who used cunning and often cruel strategies to get the land. The agents were reported to have committed such acts as murdering men in line for inheritance of the land and "discovering" long-lost heirs to the land who were willing to sell it. In 1906 oil companies began drilling in the region. Virtually overnight, the Haustecs saw the oil companies transform their beautiful land into filthy oil camps crowded with overworked men. The oil boom in the region did not last long, however, because after World War I demand for oil dropped and the oil fields became flooded with saltwater, which ruined them. Also, oilmen found enormous oil deposits in Venezuela. With the oil companies nearly gone, the Haustecs returned to their land to find it covered in oil, and the streams polluted. The Haustecs did not receive assistance from the Mexican government to rebuild and restore their community.[2] The case of Mexico is unique in that it was one of the first in a history of human rights violations in connection to oil.

Like the Haustecs, local communities in oil-producing areas around the world have voiced complaints about the deterioration of their social and economic situations. They complain that the arrival of the industry in their region (or country as a whole) caused the disruption of their traditional lifestyle and value system, encouraged the colluding of corruption leaders and the oil companies, inflamed racial discrimination, introduced an obsession for the accumulation of material wealth, caused rural-to-urban migration, escalated the cases of theft and immoral behavior, reduced the quality of life, and prompted the seizure of valuable tracts of land for oil exploration.[3] In each situation, community members believed that their rights had been violated by not only the government, but also the oil companies operating there (state-owned or private). From the governments' and the oil companies' perspective, no violations of human rights took place. Why is there disagreement over human rights—something as humans we think we inherently understand? The answer lies in a closer analysis of the meaning of the term *human rights* and how these rights are enforced.

The term *human rights* has a deceptive level of complexity to it. As humans we accept that there are certain unwritten rights awarded to us as members of human society. People expect that they hold the freedom to pursue personal goals and to be treated justly by others. This understanding is as old as human existence itself. Today, much of what makes up human rights pertains to how we are treated in a court of law and, most important, by our government. We hope that our political and legal guarantees include the freedom of speech, the right to vote, the right to privacy, and the right to due process. The problem is that while people may expect these guarantees, they are not always enforced by government. As mentioned earlier, many countries do not have clear-cut laws regarding human rights and, even if they do, no way to enforce them. Governments, without a checks and balances system in place, do not always respect the human rights of their peoples. Corrupt political leaders and ruthless dictators feel they are above the law. An international body to make governments accountable for their actions did not exist until after WWII.

Until recently, enforcement and guarantees of human rights fell under a national government's control, without an established overarching international standard. This meant that if the government itself committed violations against its own people, there was no international institution to stop it. Toward the end of World War II, in the wake of the Holocaust, the United States and the triumphant European countries created the United Nations to promote peace and ensure human rights throughout the world. In 1948, the United Nations adopted the Universal Declaration of Human Rights, which laid out the fundamental rights each person must have, and established a law making human rights violations punishable in court. From that point forward, the public has become increasingly less tolerant of human rights violations. In the case of oil, this is evident in the increasing amount of protests, strikes, and sabotages occurring within the past few decades.

In the major oil-producing countries, human rights violations have come in many forms, unique to each country and situation, but a few general patterns have emerged. The first general case, which we refer to as *community displacement*, involves the lack of government support for its people when oil companies have violated their human rights, as in the case of the Ogoni people in Nigeria during the 1990s. The second general case that occurs pertains to human rights violations in the workplace. These violations involve an oil company's hiring practices and safety standards as well as a worker's wages and benefits.

COMMUNITY DISPLACEMENT

One of the most common human rights violations has been the lack of a government's support for its people when oil companies have violated the people's human rights, as in the case of Mexico and the Haustecs. A more contemporary case occurred when the Nigerian government in the 1990s not only ignored the local community's plea for help, but also responded with brutal force toward the people who stood in the way of Royal Dutch/Shell's operations.

From the 1990s to the present, human rights abuses and environmental destruction in Nigeria related to the oil industry have become issues of major international concern. These issues were not previously nonexistent, but the conscious efforts of Nigerian political activists during the 1990s brought the issues to the attention of the United States and Europe. Consequently, foreign oil companies such as Royal Dutch/Shell have come under attack for environmental destruction and the breaking up of local communities. The focus within the international community has been primarily on the Ogoni people in the Niger Delta. However, many other groups within the region have also voiced their frustrations.

Environmental and human rights for the Nigerian people, particularly in the Niger Delta, cannot be separated, because so much of their livelihood depends on the fertility of the land. The people are rightfully furious when the land they farm and the water they fish is devastated by oil spillage and pollution. To compound this problem, development typically has been slow and the standard of living low. The people in the Niger Delta attributed this poor standard of living to being marginalized by the central government while receiving virtually none of the oil wealth that came from their region. The Ogoni, for example, pressed for separate representation within Nigeria's political system based on being a distinct ethnic group with its own language and culture. With ethnic tension already existing between those from the Niger Delta and the Nigerian government, the presence of the oil industry only exacerbated the situation.

The situation that brought the environmental and social impact of the petroleum industry onto the international scene was the case of the Ogoni people in the Niger Delta. The Ogoni are a small ethnic group of about half a million people who live close to Port Harcourt. From the 406 square miles that they claim, oil companies produce over half of Nigeria's total crude oil. For several years, the Ogoni had asked for fair distribution of oil revenues and the prevention of further environmental pollution. They complain that despite the wealth extracted from their land, they remain a poor community. They publicly denounced the Nigerian government for not enforcing the country's environmental laws.

Several of the Ogoni people took political action to voice their complaints through a group called the Movement for the Survival of the Ogoni People (MOSOP). Ken Saro-Wiwa, a writer and activist, served as the spokesperson for MOSOP and the Ogoni people, whose plight reached the international community through his rallying of sympathizers such as Greenpeace. One of the major arguments used by Saro-Wiwa and MOSOP against Royal Dutch/Shell was that the environmental damage causing the destruction of the Ogoni people's land should be considered an act of genocide.[4]

One of the most controversial aspects of Saro-Wiwa's activism was that he was said to have condoned the sabotage of Royal Dutch/Shell's oil facilities by the Ogoni people. Royal Dutch/Shell blamed the Ogoni people's protests for the loss of 3.5 million bbl of crude oil in 1992 in addition to millions of dollars worth of equipment. In response to the loss of revenue, Royal Dutch/Shell allegedly colluded with Nigeria's military by providing weapons and paying them to take action against the Ogoni. Royal Dutch/Shell denied all such allegations until several years later, when the company admitted to making direct payments to the Nigerian military in 1993. Of course, oil companies have the right to protect their industry against sabotage and turn to the government to act as mediator if necessary, particularly since the 1975 Petroleum Production and Distribution (Anti-Sabotage) Decree No. 35 empowered the military to execute anyone preventing the procurement or distribution of petroleum. However, most protests by the Ogoni were peaceful. Also, faulty equipment and accidents have often been attributed to sabotage when, in fact, sabotage was not the reason for these problems. Since oil companies such as Royal Dutch/Shell acquired a reputation as severe prosecutors, many people felt afraid to report a problem if they saw a leak.

When Royal Dutch/Shell turned to Nigeria's brutal military leader, General Sani Abacha, for help, repression became even more severe. The Nigerian government unashamedly sided with Royal Dutch/Shell and made enemies with the people from the Niger Delta. The military wrongfully arrested and detained activists, particularly members of MOSOP. Abacha used a protest rally in 1994 as a pretext for imprisoning Saro-Wiwa and eight other MOSOP leaders (later dubbed the Ogoni Nine). Allegedly, four Ogoni politicians, who disagreed with Saro-Wiwa's activism, were killed at the rally by the Ogoni Nine. The nine men were held without charge from mid-1994 to June 1995. In the fall of 1995, they were tried in a special tribunal appointed by the military government. The procedures blatantly violated international standards of law. The men were convicted, and hanged in Port Harcourt on November 10, 1995.

The environmental destruction and blatant human rights violations that occurred under the jurisdiction of Nigeria's government and its oil industry galvanized international groups such as the Nigeria-based ERA, Human Rights Watch, and Greenpeace. As protests expanded beyond the grassroots level, the World Bank confronted the issue and published its findings and recommendations. Saro-Wiwa and MOSOP became known and supported internationally. For his work, Saro-Wiwa was nominated for the Nobel Peace Prize. Nonetheless, the judicial system within Nigeria and the international community did not prevent the executions. Because of this scandal, Nigerians lost confidence in their country and publicly spoke of their country's government with contempt. It proved that the Nigerian government cared more about its oil business than its people. Furthermore, it fueled the feeling that the ethnic majority within the government and military treated the minority in the Niger Delta with disregard and contempt.

The case of the Ogoni Nine is not an isolated incident. Since then, many more groups have stood up and demanded that the oil companies respect the local populations' human rights or leave. For instance, since 1996 Ijaw youths in the Niger Delta have resorted to a permanent state of resistance by forming the Ijaw Youth Council and demanding the withdrawal of the Nigerian military and ChevronTexaco from their community. Protests in 1998 and 1999 reveal that although Abacha's reign has passed, nonviolent protests continue to be met with state violence. Instances of the Nigerian security forces killing protestors, raping women, and burning down houses have been reported.

Several solutions have been proposed to improve the human rights situation in Nigeria, and efforts have been made by both the oil companies and the government in recent years. Oil companies have proposed stricter adherence to safety standards and investment in local development projects. Oil companies in Nigeria could further reduce the impact of their operations by monitoring their facilities more closely and raising the standards for environmental protection. Oil leaks are more prevalent in

Nigeria than elsewhere because much of the equipment owned by the government needs repair. By the 1980s, oil leaks had become a routine occurrence because the government ignored early warning signs of environmental degradation given by scientists and technicians. Also oil spills could be avoided or contained better in the future. One solution that did not please environmentalists was the oil companies' decision to shift their oil operations offshore into the Gulf of Guinea. While this may appear to cause less damage to the environment and not immediately affect nearby communities, this step will only delay the necessary development of more efficient ways to collect, transport, and store oil while preventing pollution.

The government has shown an increasing concern for the welfare of its citizens in the Niger Delta since the end of the Abacha regime. During the late 1990s, the Nigerian government launched committees to inquire into the socioeconomic and environmental problems in the oil-producing regions. For example, the Niger Delta Development Panel and the Oil Mineral Producing Areas Development Commission (OMPADEC) were formed. OMPADEC was intended to survey and distribute the oil wealth to the oil-producing states and address problems of pollution.[5] Unfortunately, rumors of mismanagement and fraud have circulated about the commission. The same sorts of unsolved problems also occur for Nigeria's oil workers.

WORKERS' RIGHTS

Unfair and unsafe employment practices encompass some of the most obvious human rights violations. In the producer states, oil has created employment opportunities in the upstream sector (exploration and production) as well as in the downstream sector (refining and transportation). Most of the workers' rights violations that reach international attention deal with production and refining because of the dangers involved. The oil workers in oil-producing countries are currently employed either by a foreign- or state-owned company. They are generally the best-trained and educated workers out of all employees in industrial jobs in the country. Almost all of these workers are literate and multilingual, with English as one of the languages. These workers often operate the most sophisticated technology in the world. A great deal of responsibility lies on their shoulders because oil is a volatile fluid that flows constantly. Any accident that occurs or temporarily halts production affects a country's total oil industry. Because oil-rich countries rely so heavily on their oil industries, the oil workers hold a great deal of power. Given their importance, it is surprising how often human rights violations pertaining to wages, safety conditions, hiring practices, and benefits occur.

Oil workers are usually members of an oil workers' union. Oil workers in Nigeria, for example, can join a union that was organized by major oil companies between 1977 and 1978. These unions are part of the corporation and operate on a regional, national, and international level. In Nigeria, the oil workers' unions are divided into senior and junior staff, with the Petroleum and Natural Gas Senior Staff Association of Nigeria (PENGASSAN) for the former and the National Union of Petroleum and Natural Gas Workers (NUPENG) for the junior staff. Oil workers' unions all over the world are designed to maintain an open and positive relationship between companies and the workers as well as ensuring that the basic standards of fair pay, safe working conditions, and equal treatment among workers exist. However, these unions have little power to effect change.

When foreign oil companies had absolute control over the upstream and downstream sectors of an oil-producing country's industry, they rarely hired locals. These oil companies brought in foreigners who lived and worked in isolated camps within host countries. Since the late 1960s, however, many national governments have required that the oil companies hire local workers. Workers still complain, however, about unsafe conditions and low wages and are never hired for any decision-making positions. Until the 1970s, British Petroleum had a policy of hiring only British people. Oil companies such as British Petroleum also firmly believed that the oil business was no place for a woman and that women should not be hired unless they possessed exceptional abilities.[6] Although oil companies did eventually train and hire locals, this effort was seen by the local communities as token gesture more than serious commitment.

In several oil-producing countries, the government felt compelled to implement hiring laws to ensure that the foreign oil companies hired locally. In 1969 the Nigerian government passed a decree that stipulated the percentage of Nigerians that had to be employed by a concession holder in managerial and professional positions. The law called for the workforce to be at least 75 percent Nigerian within ten years. Furthermore, the skilled, semiskilled, and unskilled workers employed by the concession holder must also be Nigerian. Since the original policy statement was made, some oil companies, particularly Royal Dutch/Shell, have made progress, particularly in the nontechnical and semitechnical fields. In terms of key senior decision-making positions, however, little has changed. This is the case not just in Nigeria, but in most developing oil-rich countries.

Oil-producing countries with production or exploration sites today believe that their citizens should have first priority when oil companies hire employees. The oil companies see it differently. The companies in general still hire few locals and have little contact with the communities in which they operate. They feel that giving locals high-powered positions would give them influence over the company's operations as a whole. The locals that are recruited often receive security-related jobs, which keep them virtually outside the company. At the same time, however, the industry is not labor-intensive. The local people who are employed full-time receive high wages for their skilled work, but they represent a small minority and are not necessarily from the oil-producing communities. The problem with wages in an oil company, particularly in a developing country, is that they are highly subject to the political stability of the host country and the earnings of the oil company itself.

In the case of Iraq, the oil workers have endured an industry that has been highly unstable. Since nationalization in 1972, the Iraqi National Oil Company (INOC) has operated the country's oil industry. Within the INOC, the Northern Oil Company and the Southern Oil Company (SOC) have managed production operations in their respective parts of the country. The SOC is the larger and more lucrative of the two because it operates one of Iraq's largest oil fields, Rumaila. Since the removal of Saddam Hussein in March 2003, oil has represented the economic lifeline for the country; economic sanctions have been lifted and the oil industry provides the new government with the most revenue to rebuild the country. Between May 2003 and June 2004, the country was ruled by an interim government, known as the Iraqi Governing Council, under the Coalition Provisional Authority, led by the United States and the United Kingdom. As part of the interim government's rebuilding projects, it formed the Occupation Administration (OA), which assessed working conditions for state employees. The organization decided to form a new, lower wage scale based on market prices of petroleum products, foodstuffs, and the level of risk involved in the job. Oil workers disapproved of the OA's decision to also remove pay additions based on family size and location of work.

While these new wages affected all oil workers in Iraq, they affected those working for the SOC the most. Due to repeated attacks damaging 60 percent of the oil fields and facilities in the northern region since the 1990s, only the southern fields, operated by the SOC, continued to produce oil for the country. The SOC union workers responded to the new wage scale by calling for armed resistance if their demands were not met. Iraq's oil minister negotiated with the union representatives and the threat of a strike ended. In February 2004, the SOC submitted a new wage table, which included a raise in pay as well as an increase in pay based on risk and job location. The situation in post-Saddam Iraq, unfortunately, is not unique. Oil workers all over the world face the problem of potential wage cuts and a disconnection between the risk and the reward for their work.

The oil industry, of course, is a highly dangerous place to work. The high risk of the occupation spares no one—foreign or indigenous—in the oil-producing region. Although there are unions and organizations that aim to protect workers, they cannot entirely remove the risk of handling on a daily basis such a highly flammable and volatile substance. As in any industry, accidents occur. In the case of the petroleum industry, however, the accidents are often quite serious, resulting in the explosion of equipment and the loss of oil workers' lives. All phases of the oil and natural gas industry present the danger of fire and explosion. Handling of extremely heavy equipment such as pipe also puts workers' health at risk. The extreme noise of the industry can impair workers' hearing. In Iraq's Rumaila field, oil workers risked exposure to uranium waste dumped into the region during both gulf wars. Much of the technology today, especially at the refineries, is computerized, which removes much of the risk of human error; instead, the human error comes from not conducting regular safety inspections and maintenance checks. In the case of Norway, for example, an oil rig at the major Ekofisk field in the North Sea broke apart and capsized, killing 123 people. More frequent maintenance checks of the oil rig may have prevented the accident from taking place. Norway is a politically stable place; what happens to oil workers in unstable parts of the world?

Political instability and outright war threaten the safety of oil workers. During times of war a country heavily depends on its workers to produce oil, thus generating oil revenues to fund the military struggle. More often than not, innocent bystanders, in this case the oil workers, are harmed. In the case of Nigeria, ethnic conflict within the Niger Delta has made work in the oil-rich swamps highly dangerous. Ongoing clashes between the Ijaw and the Itsekiri youth escalated in 2003 and 2004. Ijaw militants have attacked ChevronTexaco's workers and facilities, demanding compensation for environmental destruction caused by the industry. In 2003, British and Scottish oil workers were kidnapped by Ijaw youth, and in 2004 workers contracted by ChevronTexaco were attacked and killed while traveling in a boat through the Niger Delta. Until the ethnic conflict is resolved in the Niger Delta, work within the oil-rich region will remain exceptionally dangerous for oil employees.

Iraq's oil workers also face great risks when they go to work in the oil fields. Since the 1980s, Iraq has been a hotbed of war, which has not only severely crippled Iraq's oil industry, but also harmed its workers. While they have not been a direct target, as in the case with Nigeria, oil workers in Iraq have received the same fate. As discussed in Chapter 5, oil fields and facilities are often a strategic target during war because their destruction can cripple a country's economy. During the invasion of Iraq to remove Saddam Hussein, the United States directed its military aggression at Iraq's oil industry. In 2002, United States and British warplanes fired at the SOC's facilities, which killed or injured several employees.[7] The Iraqi oil workers have endured a great deal of risk in doing their jobs. Because of the high risk of working in the oil business, workers have fervently lobbied for benefits to aid them and their families.

Workers in general demand certain basic benefits, which include workers' compensation when an accident occurs, a reasonable pension plan, health coverage in countries where no national health care program exists, adequate safety gear, and implementation of safety practices on job sites. What is unique about the oil industry is that workers at oil fields are often in remote regions, far away from their families. In the case of offshore oil fields, for example, oil workers stay on the platforms in the middle of a sea for weeks at a time. Also, because of the strenuousness of the work, workers often retire earlier than the national average due to health problems. In the case of Norway, oil workers sought to improve their benefits package to better reflect the peculiarities of the industry. In 2000, the workers demanded that the age of eligibility for pensions be reduced from between sixty-two and sixty-seven years of age to between fifty-seven and sixty-two years of age because of how physically demanding the work is. In June 2004, oil workers launched a strike. The Federation of Norwegian Oil Workers struck for more than a week over pensions and job security. During the strike, Norway lost the revenue from nearly 400,000 bpd. Norwegian oil workers earn $70,000 per year. They also receive four weeks vacation time for every two weeks of work. The Norwegian government responded to the strike by calling for a lockout, a legally questionable move. After several weeks, the government ordered the strikers to go back to work and directed that negotiations begin. The results of the strike have yet to be seen, but overall the government tends to respond positively to their workers given the legacy of the Labor Party. Norway's oil employees not only are among the best paid in the world, but also are said to work under the safest and most environmentally conscious conditions.

Within recent years, oil workers and environmentalists have forged a seemingly unlikely alliance. Environmentalists, of course, call for oil industries to behave in more environmentally conscious ways by reducing the toxins released into the ground, water, and air. Oil workers agree with many of the environmentalists' ideas because these policies make their working conditions safer. After all, the environment within an industry is generally a microrepresentation of the outside world. It used to be that labor unions and environmentalists stood at opposite ends of the political spectrum. Environmentalists attacked major industries, thus criticizing and threatening the oil workers' jobs. Eco-activists often formed physical barriers, preventing workers from doing their jobs, and committed acts of sabotage. Policies that regulate business, shrink large industries, protect the environment, reduce toxic waste, and clean polluted areas all potentially harm workers economically. Cleanup programs cost a great deal of money, which could come from workers' wages before it would come from a cut in the company's profits. While these groups appeared to be in perpetual conflict, economists argue that they should not be. The popular belief, however, continues to be that the higher the standards, the more workers who will be out of work. Researchers argue that strict environmental regulations and a safe workplace would actually create more jobs than they eliminate. Job security, of course, is a sensitive issue that workers do not take lightly, and any corporate change within an industry is often seen as a potential threat to jobs.

In addition to dealing with health risks associated with the job, oil workers also have to deal with the security of their jobs. As workers of the largest, most competitive companies in the world, they are keenly aware of how quickly their employers respond to expansions and contractions in the global economy. To maintain profits for their stockholders, these companies are constantly assessing their employee payrolls and competitors. As a result, employee layoffs, corporate mergers, privatization, and changes in management or ownership occur. Within the past five years, oil workers all over the world have felt the effects of several mergers between the world's top oil companies. The wave of mergers began with the merger of British Petroleum and Amoco in 1998, Exxon and Mobil in 1999, then Total, Fina, and Elf in 2000, followed by Chevron and Texaco in 2001. Each merger results in the potential loss of thousands of jobs. For example, when British Petroleum and Amoco merged in 1998, more than five thousand jobs were lost. And the major companies were not the only ones cutting oil workers' jobs.

Privatization plans within national oil companies have also caused the loss of jobs for oil workers. These state-owned companies are scrambling to become competitive on the international market. They have done this by inviting in consulting firms and foreign investors with plans of divesting, or privatizing, portions of their industry. In Nigeria, the NNPC's first privatization effort has been the divestiture of its four ailing refineries. For years the refineries barely operated at capacity, if they functioned at all. At the decision to divest in 2001, Nigeria's two oil workers' unions, NUPENG and PENGASSAN, voiced strong opposition out of a fear of their members losing their jobs. A situation such as this is a double-edged sword. If something drastic, such as privatization, is not done then these workers will be out of a job regardless. As expected, neither the government nor the oil companies want this major loss of oil revenue to continue. Both agree that the situation in Nigeria needs to improve; however, they do not always agree on how much each side should sacrifice.

While oil workers as a group respond to unfair conditions and threats to job security through strikes, they are often not alone. These workers are often part of a larger social movement. Within oil-producing regions, a mutual support system develops between oil workers and their local communities. At the same time, oil workers live and are part of the local community and are not immune to general social unrest.

RESPONSE TO HUMAN RIGHTS VIOLATIONS

Social movements all over the world develop in response to particular economic and social developments. The development of an oil industry within a country often dramatically changes not only the community's physical but also economic and social environment. Societies deal with displacement, an influx of oil wealth, and the arrival of foreign goods and culture. Of course, the degree to which this occurs varies greatly. The arrival of an oil industry in Norway, for example, did not have the same impact that it did in Saudi Arabia. This is not only because of Norway's cultural similarities with the foreign oil companies that explored its oil fields, but also because the Norwegians did not deal with the companies in the same way that many developing countries did. As discussed in previous chapters, in the late nineteenth century and first half of the twentieth century, the major foreign oil companies developed a reputation of bringing their religion, culture, and business philosophy with them to their overseas business ventures. Not all societies embraced these changes or adapted to them easily. One of the most profound examples of this is the recent rise of Islamism.

Islamism revolves around the advancement of Islam not just as a religious component to the lives of most Middle Easterners, but also as a way of life. Scholars view Islam as a comprehensive religion that provides guidelines for people to shape society but lacks clear political and economic principles. It instills the value of sharing and disapproves of greed, and thus goes against the formation of a leisure class that hoards wealth. The collecting of rentier wealth through an oil industry challenges these values, creating a conflict of interest within Islamic countries.

The emergence of Islamism represented the Middle East's response to economic and social changes largely related to the development of its oil industry. These changes began with the arrival of the Western oil companies, who brought not only a new culture, but also a superior level of technology and economic organization. The oil companies formed their own modernized enclave. The effects of this behavior destabilized surrounding traditional societies because it created a small group of elite locals who adopted a Western way of life. The majority of people, on the other hand, largely lacked a formal education and rejected this new lifestyle. Thus, a social rift occurred with underlying economic reasons.[8]

This movement followed the oil boom of the 1970s and early 1980s largely because of the government's inability to maintain the same level of wealth after oil prices declined. The aftereffects of the oil boom created social unrest and economic frustration. Much of this resentment turned into bitterness against the West especially among those who do not directly benefit from the oil industry. Thus, Islamism gained strength as the social solution.

Islamism represents only one response—and one that does not necessarily involve direct confrontation with the oil companies. More severe situations throughout the world, however, demanded more drastic measures, which included mass demonstrations and sabotage. These forms of social expression were done with the intention of grabbing international attention and causing financial harm to the oil companies, and, more important, to the host government that depended on them. In Iraq, Islamism contributed to an already dire political situation. This frustration dramatically played out in severe acts of sabotage that struck at the base of Western technology and at the government's wealth, which came from its oil industry.

Between the fall of Saddam Hussein and June 2004, Iraq was ruled by the Iraqi Governing Council. Its main purpose was to oversee the creation of a congress and prepare for national elections after the removal of Saddam Hussein from power in March 2003. The presence of Western military forces and companies such as Halliburton that were hired to rebuild Iraq (oil industry included) coupled with overall political and economic disenchantment created an increase in destruction and sabotage within the country. It was believed that frustrated oil workers played a prominent role in setting fire to strategic pipelines and storage depots, which burned for hours, wasting thousands of barrels of crude oil. When the interim prime minister, Iyad Allawi, came to power, these attacks did not cease. Oil workers in July 2004 became increasingly frustrated with the presence of Halliburton in their facilities and fields. The prime minister reported that saboteurs since December 2003 have attacked Iraq's pipelines more than 100 times. Iraq represents an unfortunate situation in which political action taken by oil workers only moves Iraq's economic situation from bad to worse.

A similar situation took place in Venezuela, only without the acts of sabotage. As in Iraq, oil workers used their leverage as engineers of the country's revenue lifeline to draw international attention to their situation and harm the government. Oil workers, backed by the opposition party, shut down the country's oil industry almost completely in December 2002 as a way to pressure the government into holding early elections in order to remove President Hugo Chávez. The protestors argued that the president was violating the people's human rights by deliberately imposing economic policies that impoverished the nation.

The opposition called for a general strike to demand that Chávez resign or hold early elections. Chávez declared the strikers' activities unconstitutional and called in the Venezuelan national guard to protect the oil industry from demonstrators. The national guard used violence, firing tear gas and rubber bullets on the protestors and injuring several people. Eight days after the strike began, Chávez fired a total of roughly 17,000 oil workers. Overall, production in the country dropped to 600,000 bpd from its pre-strike levels of about 2 million bpd. By February 2004 the opposition had collected 3.4 million signatures in a petition to force the referendum on the president. In August 2004, election officials allowed Venezuelans to vote on whether Chávez should complete the remaining two and a half years of his term or not. The results favored Chávez, which raised accusations of fraud.[9]

While protests and acts of sabotage are deliberately planned ways of attracting public attention, groups that feel that they have been oppressed by the destruction and practices of the oil industry in their communities have also teamed up with international human rights organizations to get their message heard throughout the world.

HUMAN RIGHTS ORGANIZATIONS

Most human rights organizations take on a broad range of cases, from the oppression of women and children to the unjust imprisonment of activists. Since they are nongovernmental organizations that do not work for profit, they are not bound by ties of politics. Their primary focus is to draw international attention to these abuses and use publicity as a tactic to sway the public to place pressure on governments or industries. When human rights abuse has occurred, generally the victim or people campaigning on their behalf seek out these organizations. In turn, the organization will send its own team of researchers to conduct studies, often scientific, on the situation. Most members of these organizations work on a volunteer basis. With the information gathered, they can assist the victim in taking legal recourse at an international level.

In 1948, the United Nations adopted the Universal Declaration of Human Rights, which clearly states that everyone has the "right to life, liberty, and the security of person." Under international law, any violation of these rights can be tried in a court of law. Human rights organizations use the Declaration of Human Rights as their most powerful tool against rights abusers. While there are a multitude of human rights organizations, there are a few notable ones that have dealt with cases of human rights abuses related to oil.

Amnesty International

The largest and most active organization, Amnesty International stands out for its success and commitment to the protection of human rights. The organization began in 1961 with a British lawyer writing an editorial to the *Observer* regarding a human rights violation. Within six months, this brief publicity effort was being developed into a permanent international movement. This sparked the beginning of the organization. Today, Amnesty International is a phenomenal organization that campaigns for human rights all over the world. Much of its work related to oil involves the violations of human rights of local oil workers by oil companies. In addition to its work on the detention of two oil union members in Nigeria during the 1990s, it is also heavily involved with the role of oil in the Sudan as well as in Colombia.

EarthRights International

EarthRights International (ERI) is an organization that emphasizes the connection between environmental and human rights. Through international support and legal action, ERI works toward the reduction of environmental and human rights abuses. The organization boasts two headquarters: one in the United States and another in Thailand. Much of its work has focused on the human rights violations and environmental destruction caused by oil production in Nigeria, Burma, and Ecuador.

Human Rights Watch

Based in the United States, Human Rights Watch has worked for the preservation of human rights throughout the world since 1978. The organization originally formed in Helsinki to monitor the Soviet bloc's commitment to the Helsinki Accords. Shortly after its formation, it took on a truly global cause by expanding its operations to the Americas and Africa. During the 1990s, it became a vocal opponent to the human rights atrocities against the Ogoni people living in the oil-producing region of Nigeria.

CHARITY AND THE OIL COMPANIES

Thus far we have focused primarily on human rights abuses and the political actions taken to remedy these problems. Not all developments having to do with human rights are negative, however. For all of the social and economic unrest caused by the arrival of foreign oil companies in the late nineteenth century and the twentieth century, much that is good has come from their presence. And this trend continued after nationalization through state-owned oil companies. Independent and state-owned oil companies have voluntarily spent billions of dollars on development projects. These companies often went beyond the benefits demanded by the oil-producing country. They sought to provide basic amenities such as paved roads, electricity lines, and sanitary living conditions for workers. In many cases they developed homemaking courses for women and oil industry training programs. They also built schools and hospitals in rural areas. In some cases, the independent oil companies spent more on developing a community than the national government did.

During the early twentieth century, philanthropy was at a high point in the United States, and it spread through the foreign oil companies as well. While wealthy individuals such as Andrew Carnegie shared their wealth throughout the United States, Nelson Rockefeller of Standard Oil spread his wealth in oil-producing countries such as Venezuela. A part of their investment was motivated by the logic that it was more profitable for a company to have healthy workers than ill ones. Therefore, it was in their best interest to improve living conditions and health services. Also, these companies recognized the importance of creating an agreeable and stable relationship with the national government if they planned to continue their operations. No company during the early twentieth century understood this more than Standard Oil.

In Venezuela, Nelson Rockefeller, the grandson of John D. Rockefeller (the creator of Standard Oil) created the Creole Foundation. At the time, Creole Oil was the Venezuelan subsidiary to Standard Oil. The idea for the organization began in 1937, when Nelson Rockefeller visited oil workers' camps and was devastated by the living conditions. He saw workers living as squatters when they were supposed to have living quarters. But it was not just how the workers lived that moved him; he saw people in the surrounding communities living in poverty as well. After taking the situation up with Venezuela's president at the time, Eleazar López Contreras, Nelson Rockefeller focused his philanthropic ideas on Venezuela's public sector. The Creole Foundation contributed funds through other organizations to develop Venezuela. By the late 1950s, Rockefeller's contribution totaled over $5 million.[10] The charitable contributions of the Creole Foundation created a legacy. They reminded oil companies—independent or national— all over the world of the great potential of social investment.

THE FUTURE OF HUMAN RIGHTS?

Two seemingly contradictory trends are emerging with regard to human rights. On the one hand, concern over human rights has significantly increased. On the other, human rights violations in relation to oil development appear to be on the rise. With the expansion of the Internet and other forms of global communication, the spread of information about abuses and an international response to them happen much more quickly than ever before. The invention of the Internet makes information such as this available to almost everyone in the world. For this reason, countries and companies have had to become more effective in either covering up their actions or resolving the situation.

In the case of Nigeria, when people protested against the presence of Royal Dutch/Shell and ChevronTexaco in their communities, the oil companies responded by spending more money on improvement projects than the other oil companies did. In 1996, Royal Dutch/Shell claimed to have spent more than $36 million. It also launched a scholarship program, which cost more than $21 million a year. In 2004, the Nigerian parliament called for Royal Dutch/Shell's Nigerian subsidiary, Shell Petroleum Development Company, to pay the Ijaw people $1.5 billion in compensation for allegedly causing health problems. Royal Dutch/Shell admitted that its activities partially contributed to conflict in the Nigeria Delta.[11] Recently, ChevronTexaco found itself in the middle of a rivalry between communities in the region because it hired more local people from one group than from another. It also invested in development projects to appease one group of protestors, which prompted another group of protestors to demand the same benefits. Having learned from Royal Dutch/Shell's mistakes, ChevronTexaco has committed to community development and may be giving more within the next few years. It is evident from these examples that development projects within Nigeria as a whole need to be designed and allocated with great care in order to minimize tension. And foreign oil companies are not the only ones stepping up to protect human rights.

Within the past few years, more and more national governments have taken an active interest in human rights because they did not want a repeat of the twentieth century, during which foreign oil companies operated virtually unfettered, without regard for human rights. While oil-rich nations have privatized their national companies in order to encourage investment, they are making sure to draft legislation that protects people and land. Russia, for example, nationalized its oil industry in 1991, but has factored in a protection plan. In Russia's 1993 constitution, an article guaranteed local people rights within the federation. The government signed agreements with the oil companies, stating that the government reserved the right to oversee drilling operations in an attempt to protect the indigenous peoples of the Khanty-Mansi Autonomous Area in the heart of Russia's oil-producing region. Russia's actions may indicate a new wave of social consciousness.

Human rights organizations all over the world are hoping that the 1990s marked the beginning of a new relationship between international oil interests and the protection of people. Perhaps companies and governments now realize the long-term economic and political costs of human rights abuses towards indigenous communities and oil workers. The risk of living near oil fields and facilities will never change, nor will the risks involved in working within the industry. Companies and governments together, however, can work to reduce the impact on the people and the land.

FURTHER READING

Human rights have become a major field of research within the past thirty years. From this growing awareness a number of excellent works that deal not only with human rights from a theoretical perspective, but also with case studies throughout the world, have emerged. For works that focus on legal and corporate responsibility in regard to human rights, see Fergus MacKay's "The Rights of Indigenous Peoples in International Law" and Lyuba Zarsky's "Global Reach: Human Rights and Environment in the Framework of Corporate Accountability" in Lyuba Zarsky's *Human Rights and the Environment* (Sterling, VA: Earthscan, 2002).

For information on Amnesty International, visit its Web site (www.amnesty. org), and see the annual *Amnesty International Report*, which highlights human rights abuses around the world as well as work in progress to resolve them. For information on the Human Rights Watch organization, see its Web site (www.hrw.org). For more information on EarthRights International, visit its Web site (www.earthrights.org). For more information on human rights groups, see Amnesty International's *Nigeria, Repression of Women's Protests in Oil-Producing Delta Region* (London: International Secretariat, 2003). For an excellent documentary on the 2002 attempted coup in Venezuela, see Kim Bartley and Donnacha O'Brian's *This Revolution Will Not Be Televised* (Dublin: Irish Film Board, 2002). For more on the case of Nigeria, see Browen Manby's *The Price of Oil: Corporate Responsibility and Human Rights Violations in Nigeria's Oil Producing Communities* (New York: Human Rights Watch, 1999) and Ken Saro-Wiwa's *Genocide in Nigeria: The Ogoni Tragedy* (London: Saros International, 1992). For more on the case of Mexico's oil workers prior to expropriation, see Jonathan Brown's *Oil and Revolution in Mexico* (Berkeley: University of California Press, 1993). For a discussion of the 2003 strikes by women in Nigeria, see Terisa Turner and Leigh S. Brownhill's *Why Women Are at War with Chevron: Nigerian Subsistence Struggles Against the International Oil Industry* (New York: International Working Group, 2003).

Chapter 8

Oil and Economic Development

International hypocrisy has a name: Aid. It will never be a source of revenue necessary for development—that can only be sustained by just and equal terms of trade.

President Andrés Pérez, 1976[1]

In this chapter we return to national interests by examining oil-rich countries' push toward economic development. We begin by examining the second and third oil shocks and devote the remainder of the chapter to the years that followed. We focus on the post–oil shock years because they provide a window into understanding how oil-producing countries handled the abundance of oil revenue that they received during the price hike. The main question we seek to explore is how effective the countries were in allocating the revenue and establishing development programs. Did the countries spend the wealth on luxury items, the military, or health and education projects for their people? While these countries did invest in a variety of development projects, they were not, ultimately, successful in creating sustainable economic growth. In fact, after the price hike of the 1970s, many developing oil-rich nations have found themselves seeking loans from the World Bank Group, OPEC Fund, and International Monetary Fund (IMF). But how does an oil-rich country achieve sustainable growth in the face of environmental concerns, international pressures, and political instability?

Nearly every country, government, and person in the world is in favor of economic development, with an understanding that economic development leads to a better way of life for all. What exactly is this cure-all strategy called economic development that most countries want but few have achieved? *Economic development* is the sustained growth in per capita income in conjunction with the reduction of poverty as well as the expansion and diversity of the economy.[2] The problem with economic development is that it implies social change, which is not easy to implement. Furthermore, there is little agreement as to how development is achieved. In addition, economic development does not necessarily lead to the overall economic growth of a country. Oil-producing countries often receive a great deal of oil revenue, but the incomes of their populations do not necessarily improve. In most cases, it is because the oil wealth reaches either the government or the foreign or state-owned oil companies, and there it stops—never fully reaching the people.

In a discussion of development, the first question that emerges pertains to how a country spends its oil wealth. We established in earlier chapters that the global oil industry is big business. For any country, developed or developing, oil provides the most lucrative form of revenue a country can have. Although the resource is exhaustible, the short-term gains in terms of profits are tremendous. The period that most benefited oil-producing countries was that of the oil shocks of the 1970s.[3] At that time, the price of a barrel of crude oil from Saudi Arabia went from around $7 (1973) to around $25 (1974). Countries such as Venezuela, Norway, and Nigeria, which were not directly involved in the 1973 oil embargo, especially benefited because the United States relied on them as a source of oil when the Arab countries cut off supplies.

The three oil shocks occurred during the 1970s and 1980s, with the last occurring in the early 1990s. These events are called *oil shocks* because they were quick, unforeseen shifts in oil prices that rippled throughout the world, leaving no region untouched. The first oil shock took place in the early 1970s with the culmination of oil shortages, political conflict in the Middle East that led to an oil embargo in 1973, and mounting frustration among oil-producing countries toward the foreign oil companies regarding concessions. The first shock sent oil prices to never-before-seen levels. OPEC, which had previously been ineffective in controlling prices, took control of oil prices and set them at the rates it felt best for itself and the consumers. For the first time, the world realized OPEC's power and control over the world's oil. Prices began to return to a new norm around 1975. Then another price hike began in 1979, steadily increasing until 1985, again due to political instability in the Middle East. The third oil shock, which took place between 1985 and 1990, actually drove prices down. Collectively, the three oil shocks affected the world dramatically.

THE SECOND AND THIRD OIL SHOCKS

The second oil price shock started with the Iranian Revolution in 1979, during which the monarchy was overthrown and a new Islamic republic headed by Ayatollah Khomeini took control. At the time, Iran produced a high percentage of the world's oil. While production did not stop in Iran, it did slow down. Also, it created a wave of uncertainty for the international oil market. Following the Iranian Revolution came the Iran-Iraq War, which began in 1980 and lasted eight years, ending without a clear win on either side. While the war broke out due to a territorial dispute over who controlled the Shatt-el-Arab waterway, underlying tensions had been brewing for years.

Competition for regional dominance as well as political and religious differences set the two countries at odds. The war had a major impact on each country's oil industry, as fields were burned and facilities destroyed. As a consequence, oil exports from Iran and Iraq decreased (by more than 2 million bpd in the case of Iran and almost 1 million bpd in Iraq), affecting the global supply and the price of oil. Within the first few years of the Iran-Iraq War, the price of oil increased from around $13 per barrel (1978) to roughly $35 per barrel (1981). As during the first oil shock, the increase in the price of oil meant that other oil-exporting countries enjoyed an increase in oil revenue. This time, however, consumers were ready.

When the Iran-Iraq War broke out, consumers all over the world braced themselves. Not enough time had passed for oil-importing countries to effectively develop an alternative source of energy. Even though countries such as the United States vowed to reduce their reliance on Middle Eastern oil after the 1973 oil embargo, they did not succeed. Consumers responded to the crisis by buying petroleum products in excess and hoarding them for the future. By this time, the members of the OECD had set an emergency program in place that called on members to hold ninety days' worth of oil in storage. Members were also ready to assist each other to avoid shortages. Furthermore, a worldwide recession hampered the demand for oil. OPEC did not expect this reaction; it calculated incorrectly that world demand would remain the same. At the same time, OPEC members could produce less oil without cutting into their profits because of the increase in prices. OPEC members began to keep oil in the ground. The decline in demand outweighed the rate of reduced production, causing a drop in the price of oil, and thus the third oil shock began.

The third oil shock, unlike the first two, involved a dramatic drop in the price of oil. As with the first two shocks, the world was not entirely prepared and acted on instinct. By 1985, the key oil producers began to feel the crunch and decided it was time to take action to control the price of oil. Saudi Arabia played a critical role in OPEC as the swing producer. Because Saudi Arabia is the world's largest oil producer, it could (and still can) afford to increase or decrease the amount of oil it produces quickly without creating domestic economic problems. Even today, Saudi Arabia acts as a swing producer. Indeed, when prices began to fall, it was up to Saudi Arabia to stop them. OPEC had attempted to set the price of oil at around $18 per barrel but was not successful. In 1990, with the onset of the Gulf War, prices went even lower. Crude oil prices during this time dropped by roughly one-third. Because it was not cost-effective for oil companies to explore new oil fields, already existing oil fields appreciated in value. The 1994 upturn in price came from a mixture of inflation and the coordination of pricing among OPEC and non-OPEC countries such as Mexico, Norway, and Russia. After the crisis ended, assessments of how the oil-producing countries handled the oil shocks emerged from all sides.

The first two oil shocks brought an extraordinary amount of wealth in a short time to oil-producing countries. At the same time, they plunged the world into a recession. Economists agree that if it had been handled well, the oil-producing countries could have improved their economic situation by leaps and bounds. The truth, however, is that this ideal situation did not materialize. A combination of preexisting political and economic instability prevented leaders from using the money wisely.

THE USE OF PETRODOLLARS

The use of petroleum revenues, or *petrodollars*, in any oil-producing or oil-exporting country has to be considered with care. Petrodollars are quite different from other forms of income because they represent rent, which is revenue received from an industry in which the recipient collects without being directly involved. When a person rents an apartment, for example, the landlord receives money as compensation for the person living in the space. Likewise, the government receives income tax from its citizens without being directly involved in how the people earned the money. Within the oil industry, petrodollars are rent. They represent easily earned money because a government does not have to go through the pains of establishing local industry to receive hard currency. With so much money flowing into the country, the incentive to develop local industry is pushed aside. Oil revenues often become a substitute for other forms of national income instead of acting as a supplement. The problem is that this apparent boom only lasts as long as oil prices are strong and oil continues to flow. Ideally, oil-producing countries should use petrodollars to supplement the income they receive from other sources of income (i.e., local industries). Juan Pablo Pérez Alfonzo, the oil minister of Venezuela during the oil crisis, recommended using the oil wealth to improve the agricultural sector of the economy, or "sowing the oil" as he called it.[4]

Before the discovery of oil, most of the oil-rich countries in the world were primarily agrarian. With the discovery of oil, these countries ceased supporting the agriculture industry. The result has been disastrous. Little to no support has come from the government to farmers using outdated equipment. A large number of farmers either left their farms in search of urban employment or shifted increasingly toward cash-crop production. A prominent example is the case of Iraq, which had acted as the breadbasket of civilization for thousands of years, producing valuable foodstuffs. But by the 1980s, it could barely meet its own demand. Without agriculture as the economic failsafe behind oil, countries could not weather the third oil shock easily.

During the oil shocks, the heavy reliance on petrodollars and the lack of long-term planning on the part of the oil-producing countries became painfully obvious. During the first and second oil shocks, developing oil-rich countries received an influx of petrodollars virtually overnight. Instead of investing in the industrial and agricultural sectors, they often spent the money on luxury items or unsuccessful projects. For example, the price of Venezuela's crude oil between 1972 and 1974 rocketed from $2.50 to $10.50 per barrel. With this new wealth, the Venezuelan government in 1974 attempted to take half the oil income and place it in an investment fund called the Fondo de Inversiones de Venezuela. The fund was supposed to promote public investment within and outside Venezuela and an import substitution program to improve Venezuela's economy. The true function of the fund emerged shortly after its formation when the Venezuelan government began to use it to pay for public expenses.[5] Instead of spending the money on development projects, governments spent it on luxury consumer goods, including imported food, clothing, and cars, as well as on enormous conference centers and other symbols of national wealth such as the president's palace, car services, and national celebrations. Like Venezuela, Nigeria managed its petrodollars poorly, which caused economic decline. In addition to an obscene amount of imported luxury goods, particularly expensive European cars, the Nigerian government spent its petrodollars on feeding its political patronage system.

The oil boom actually reduced the standard of living for the bulk of Nigeria's population. This is because the political leadership operated along ethnic lines and with regional loyalties. As a consequence, oil revenue often went to specific regions and slighted others. During the 1973 oil embargo (October 1973 to March 1974), Nigeria's oil revenues nearly quintupled because of the high prices, increased production, and high taxes and royalties received. Nigeria was under the military rule of General Yakubu Gowon from 1966 to 1975 and Brigadiers Murtala Mohammed and Olusegun Obansanjo from 1975 to 1979. As a way to maintain power for these leaders, the oil revenues were spent on an increase in wages for the military and for civil servants. The end result of having invested the revenues this way? That money was used to fund a military junta that overthrew the Gowon government, claiming a desire to cleanse the government of corruption and poor economic policy. This, of course, was really an act of the military junta moving into a position of power to catch the oil wealth as it flowed into the country. As a result, oil wealth in Nigeria fueled political instability and economic decline. Rivalries emerged between militant groups over the available oil revenues, causing armed conflict, as autocratic leaders used oil revenues to achieve and maintain power.[6]

In addition to leaders spending petrodollars on improving the prestige of the government, they also spent a great deal on military and security forces as well as weaponry. The priority given to these purchases only indicates how tenuous the position of these leaders was. This was particularly true in the case of Nigeria and Iraq. Much of their spending turned into an almost direct exchange of oil for arms. When the oil price hike began in the 1970s, the number of weapons imported into unstable oil-producing countries increased. This mismanagement of petrodollars only strengthened the position of autocratic leaders.

The perfect example of this problem is Saddam Hussein's pursuit of political control of Iran in the 1980s. Scholars have argued that the Iran-Iraq War, and consequently the misuse of Iraq's petrodollars, stemmed from a leadership based entirely on personal interests and vendettas. It shows the extent to which Iraq's economic downfall was attributed to political corruption. Saddam Hussein's personal rivalry with Ayatollah Khomeini dated back to the 1978 expulsion of Khomeini from Iraq. It has been argued that Saddam Hussein swapped development plans using petroleum revenues for Iraq in exchange for military spending during the Iran-Iraq War. Prior to the Iran-Iraq War, Iraq had shown signs of economic diversity, with reduced reliance on its oil industry. Once wartime spending began, however, Iraq became more dependent on oil revenues than ever.[7] The situation in Iraq continued to decline, as we shall see, with subsequent conflicts in the 1990s.

The poor usage of petrodollars neither protected oil-producing countries from the second and third shocks, nor promoted economic development. To make matters worse, the third oil shock occurred before economic development projects implemented during the first two shocks reached fruition. Many of the projects were in their early stages and the governments could not collect sufficient revenue to complete them. Having spent their money unwisely, many oil-producing countries turned to international lending institutions for financial assistance during the 1990s.

INTERNATIONAL AID INSTITUTIONS

While there are several institutions providing aid, we will examine those used most by oil-producing countries: the OPEC Fund, the World Bank, and the IMF. These institutions, despite popular belief, do not work in opposition but complement each other in meeting the needs of developing countries. In Chapter 5, we mentioned briefly the emergence of developing countries engaged in a new foreign policy of assisting other developing countries. With a sense of prosperity and economic security, oil-rich countries saw that one of the best ways to assist neighboring countries was through sharing their wealth during the first two oil shocks. Since the third oil shock, these endowed countries have found themselves turning to aid institutions to get them out of financial difficulties.

OPEC FUND FOR INTERNATIONAL DEVELOPMENT

The OPEC Fund for International Development is a small development finance institution. Members of OPEC formed it in 1976 to promote financial cooperation among its member countries as well as provide assistance to other non-OPEC developing countries. It is owned by thirteen developing countries that contribute a portion of their oil sales. In 2003, the OPEC Fund granted funds for more than 600 projects worth a total of almost $300 million. The OPEC Fund is a unique organization that has changed the way people think about aid.

OPEC proved that development aid was not just a venture for rich, industrialized countries. For the first time, a group of developing countries who shared all the same development problems decided to provide assistance to other developing countries. The borrowers often feel more comfortable going to OPEC for assistance than the World Bank because of this dynamic. Initially, the OPEC Fund was seen as compensation for high petroleum prices to developing oil-importing countries because the institution formed during the oil shocks of the 1970s. In defense of this accusation, OPEC is quick to point out that although the formal OPEC Fund came into being in 1976, it had been providing aid before the first oil shock in 1973.[8] The OPEC Fund prides itself on avoiding lending based on self-interest, a fault it perceives the West as being guilty of. OPEC countries were not through the OPEC Fund trying to protect strategic locations or attempting to use financial assistance to maintain markets for their petroleum products. In fact, the OPEC Fund often finances energy projects that encourage the development of alternative energy sources that would decrease the borrowing country's consumption of oil. The OPEC Fund is not necessarily a resource in opposition to the IMF and the World Bank; in fact, OPEC members have contributed to the World Bank. Many developing countries, however, find it easier to approach the OPEC Fund for assistance than the World Bank or the IMF.

THE WORLD BANK AND THE IMF

The World Bank is an agency under the United Nations that provides advice, loans, and technological expertise to developing countries in order to reduce poverty and promote economic growth. It was formed in 1944 at the Bretton Woods Conference in Bretton Woods, New Hampshire, with forty-four representatives from around the world. Although the organization is called a bank, it is actually not a bank in the true sense of the word; instead, it is a collaboration of five different financial agencies that together offer low-interest long-term loans and grants to developing countries. The World Bank is organized like a cooperative, with its member countries as shareholders. The United States is the largest single shareholder. In 2003, the World Bank provided $18.5 billion to more than 100 countries for development projects, among which are oil and gas ventures.

Within the World Bank operates the Oil, Gas, Mining, and Chemicals Department, which is an agency dedicated to the development of energy resources. It provides developing countries with legal and technical advice regarding all aspects of establishing or enhancing an oil and gas industry. In the 1990s, after the fall of the Soviet Union, Russia received help in developing new oil fields. In 1991, the Russian oil sector opened up to foreign investment in exploration and development. One of the first companies to make a major investment in developing a Russian oil field was Conoco. In January 1992 Conoco established the Polar Lights Company (PLC) as a 50-50 joint venture with a Russian oil company to develop the Ardalin oil field northeast of Moscow. The PLC received $200 million in financing, allowing the joint venture to proceed with the production project despite the political uncertainty of the country.

The IMF came into being in 1945 at the close of World War II. While the IMF does not directly provide assistance for the expansion or improvement of a developing country's oil and gas industry, it does take an active interest in how the country handles its oil revenues. The IMF, like other specialized agencies, offers short-term lending and advice to developing countries. It attempts to prevent crises within the international monetary system by encouraging developing countries to adopt beneficial economic policies. When a country asks the IMF for financial assistance, the IMF offers the country advice on how to improve its economic situation in order to effectively use the loan. It offers recommendations to follow a strict *structural adjustment program*, which is a plan to reduce public spending, such as subsidies, privatize state companies, and increase the nation's participation in the free global economy. A large number of developing countries, particularly oil-rich ones, have gone to the IMF for assistance.

The World Bank and the IMF often collaborate in making support available to governments because each has an area of expertise that compliments the other. The IMF focuses on monetary policy while the World Bank works more on assisting countries in development. In other words, the World Bank advises on social policies, while the IMF promotes sensible economic ones. Through joint meetings, the two institutions strategize on improving the success rate of their work. Because the World Bank and the IMF work so closely they are often referred to as one institution.

Going to the World Bank and the IMF is not easy for an oil-rich developing country. One of these countries' major reasons for nationalizing in the first place was to remove the influence of wealthy Western countries. In the case of the Middle East and Africa, nationalization meant removing the last traces of colonial rule. For Latin American countries, it meant the removal of Western countries from heavy involvement and manipulation of their economies. For these countries to go back to the West admitting that since independence their political and economic situation has gotten worse and not better is quite difficult. Also, any country that goes to the World Bank and the IMF for help knows that it must implement a great deal of structural change to even be eligible for the loans.

After the second oil shock, countries such as Venezuela faced serious financial and political trouble. A parade of unfocused governments came and went during the 1980s. In 1987, Venezuela went to the IMF for assistance. The IMF recommended a "shock therapy" form of structural adjustment, which Venezuela tried to follow. The program included restoring the country's ability to pay off external debt and making the country less dependent on state funds. The results were apparently mixed, with Venezuela more dependent on oil revenue than in previous years.[9] Unfortunately, Venezuela's case is not unique. Experts on economic development have argued that the IMF's approach lacks sensitivity.

The IMF, in particular, has been criticized for handing out essentially the same cure for different diseases. It assumes that all developing countries suffer from internal mismanagement. As a result, the IMF offers virtually the same package to all borrowers: devaluation of the local currency, reduction of public spending, and decreases in wages, especially for the public sector. Furthermore, the IMF has always encouraged countries to reduce the level of government involvement. It suggests the removal of government controls on trade and the elimination of subsidies.[10]

Sometimes countries follow the IMF's recommendations without taking a loan, but in order to gain credibility with the fund to negotiate debt or attract new loans in the future. Also, many countries hope to follow the IMF's advice without its financial involvement. The most basic requirements made by the IMF are to reduce government involvement in two major ways: discontinue price subsidies on products such as fuel oils and privatize state-owned industries. Both of these changes have dramatic effects on an oil-rich country's petroleum industry.

SUBSIDIES

The use of petroleum-based subsidies is easy to advocate, difficult to sustain, and almost impossible to remove. Subsidies act to lower the price of a product, in this case oil and gas, to the buyer and thus encourage its purchase. In simple terms, a subsidy is the difference the government pays between the export and domestic price of a product. Countries that choose to subsidize do so in hopes that it will promote economic development. Petroleum subsidies are easy to advocate because they appear to be a simple solution to encouraging development. Oil-producing countries hope that making petroleum products more affordable will encourage local investment in industries. Mexico, for example, maintained this position during the 1970s. In the midst of an increase in oil reserves as well as petrodollars, Mexico's government decided that the way to carry out national development plans included the establishment of low domestic prices for petroleum products so that they could be afforded by a wider span of the population. Many people in oil-producing countries believe it is their right as citizens to have access to their country's resource. Encouraging people to use petroleum products such as kerosene for cooking reduces the destruction of forests.

Governments also use subsidies so that remote parts of the world have access to petroleum products at the same price. In the case of Nigeria, subsidies are used not only to make the cost of petroleum products low, but also to make the cost of the products uniform throughout the country. The government ends up paying for the high transportation costs of oil trucks traveling long distances or into remote parts of the country where some roads are nearly impassable. Indeed, the general populous embraced the subsidy system almost immediately because it brings with it also a sense of national pride and a belief that the government is truly working for them. One Nigerian described this perfectly in an editorial column written in 1978:

We know that a government exists for the people and a government seeks to make life worth living for the people. . . . That is why every government subsidizes so many things so that people do not suffer. A government does not think of profit all the time to the detriment of the masses.[11]

One of the biggest arguments for subsidies by the general populace is that all citizens of an oil-producing country should benefit. They view it as a crime to always choose exportation over domestic consumption. Unfortunately, it is financially difficult for a developing country to sustain this policy.

Petroleum subsidies are difficult to sustain; they are the most expensive form of government spending, aside from military spending, because the government receives little to nothing in return except temporary political satisfaction. The government uses both military and subsidy spending to win public favor or as a way to maintain power. They are both short-term solutions to rally the public behind a leader, but they do not help the country in the long run. The governments who choose to subsidize are correct in their idea that making oil and gas affordable for everyone encourages development. But they do not anticipate what impact it will have on overall production. In the case of Venezuela during the 1970s, the government never considered scarcity as an issue, but the subsidies created an increase in the quantity of petroleum products demanded within the country. In order to satisfy both export and internal demand, Venezuela either had to expand its industry or shift oil reserved for export to sell within the country, thus cutting oil revenues. Venezuela's refineries reached their limit and expansion projects had to be drafted. The unfortunate truth, however, is that a sizable percentage of the crude oil reserved for domestic consumption does not remain in the country.[12]

When one country offers petroleum at prices lower than the international market price, its neighbors take advantage of the situation. Nigeria, for example, sells petroleum products for lower prices than its neighbor Niger to the north and the Republic of Benin to the west. The result is a rampant problem of Nigerians along the borders smuggling cans of gasoline across the border to sell. A substantial portion of Nigeria's oil leaves the distribution terminals in tankers and never makes it to its intended destination. In the meantime, the government, which is subsidizing, can easily watch the international market and see what it could earn if all of its products were sold at market value.

Petroleum subsidies are difficult to remove for several reasons. First, the population never fully understands just how much it costs the government to maintain this national luxury and, therefore, does not accept the decision to halt subsidies easily. It is common for a country's population to expect its government to provide what everyone would consider a "right." When the oil boom ended in Mexico, the government quickly realized its need to discontinue subsidies, but a fear of popular reaction to price increases delayed the implementation of this policy. This kind of public reaction, however, is not just about an increase in fuel prices.

The decision to lift a subsidy usually occurs in conjunction with two other difficulties: severe economic problems and the decision to ask for foreign aid. Just when the economic situation is difficult and much of the population is unemployed or underpaid and the standard of living has deteriorated, the government makes the cost of fuel out of reach for a section of the population. People that depend on their own vehicles or public transportation for their jobs find the removal of fuel subsidies difficult to bear. The most important problem with fuel price adjustment is that it affects not just petroleum products but the cost of everything within the country. The second difficulty, asking for foreign aid, represents waving the white flag of defeat for many oil-rich developing countries. In some ways, it is an admission of failure for a government. As mentioned earlier, one of the first actions the World Bank and the IMF ask developing countries to take is reducing government spending by discontinuing the use of fuel subsidies. Oil-rich developing countries face the same problem when the World Bank and the IMF recommend that they privatize much of the public sector as well.

PRIVATIZATION

As discussed in Chapter 3, there are numerous benefits to developing a national company, but there are also major drawbacks that present roadblocks to success. After nationalization, state oil companies often found themselves grappling with a lack of capital, a lack of managerial and technological expertise, and an often restrictive relationship with their governments. These problems and many more began to surface almost immediately and became only more glaring as time passed. After nationalization, problems of corruption, failing equipment, and the inability to meet domestic needs emerged as well. All of this, of course, developed within the larger context of economic instability. By the 1990s, the oil-rich countries knew something had to be done.

Privatization occurs when the state divests itself of enterprises that it owns and operates. Methods of privatizing a state-owned company vary just as methods of nationalization do. In the case of privatizing, however, the method depends on what fraction of the state's holdings the government wants to divest. Also, governments have to decide whether they would like to sell the shares to one private company or to individual shareholders. Privatization does not just mean total divestiture, either. It can mean the introduction of private investors in the form of management contracts or simply restructuring the state company to act as a private company. Little by little, state-owned companies flirted with the idea of privatization, but governments knew it was not a decision to take lightly. All the reasons to develop a national oil company still existed. These ideas of national pride and independence did not die.

The national oil company continued to foster a sense of pride, especially for those who went through decades of colonialism and a fierce battle for independence. These nations felt that the state-owned company, ideally, was looking out for the best interests of the country by making a highly valuable resource inexpensive and available to everyone. To divest this source of pride from an oil-producing nation is not easy. This is particularly true when divestiture comes at the recommendation of the World Bank and the IMF. As with removing subsidies, the common response to taking advice from the World Bank and the IMF is the feeling that the government failed. But these reasons of pride had to be placed behind the long list of problems that emerged from having a state-owned oil company.

While the success or failure of the national oil company varied from country to country, the ultimate result thus far has been a uniform move toward privatization. Many state-owned oil companies found themselves in a quandary. They had trouble balancing production and domestic demand so that profits would not fall. The governments depended on this revenue, and any small disruption in its flow made further investment projects difficult. With this kind of instability, private lenders expressed hesitancy in extending credit. In order to attract the necessary funds, privatization became the best option. Part of the problem was that new developments within the oil industry simply cost more than before.

The feasible oil and gas fields in many of these countries have already been found, and at this point countries such as Saudi Arabia do not expect to find more sources of crude oil within their borders. In the case of Venezuela and Norway, any future oil and gas reserves exist offshore, which is the costliest form of oil production. The equipment and technological expertise for offshore production is expensive. Also, the latest trend among oil-producing countries has been the development of natural gas. This resource has quickly become popular among oil-importing countries because it is cleaner to burn and easier to transport. In the near future, natural gas reserves may easily outnumber crude oil reserves. Hoping to catch this wave, oil-producing countries are looking to develop their natural gas industries. The costs of the technological expertise, highly sophisticated equipment, and capital needed to explore and begin production are greater than most of these oil-producing countries can afford. Even Norway, which is in the best economic shape, has asked foreign oil companies to invest in its industry.

Since the late 1960s, Norway's state-owned oil company, Statoil, has had extensive control over the industry. Norway's national assembly, the Storting, claimed total ownership of Statoil. In addition to Statoil, the Storting owned the majority of Norsk Hydro, which is the second largest oil company in Norway. The oil industry ran this way until recently, when the Storting began to take the idea of privatization seriously. This decision has coincided with a decline in crude oil production in the North Sea. Within the past few years, crude oil production has fallen by almost 200,000 bpd; it will continue to decline as more and more fields become mature. In 2001, Norway approved plans to sell up to 25 percent of Statoil to private investors. In May 2002, Statoil sold a portion of its crude oil assets to a Danish oil company for almost $130 million in order to shift its focus to new developments. Also, Statoil managed the state's direct financial interest (SDFI), which operated more than 100 oil and gas fields on behalf of the state. Part of this 2001 plan was to take the SDFI and sell off shares of it on the New York and Oslo stock exchanges. By 2004 the government reduced Statoil's holding to 77 percent. With the remainder of SDFI, the storting formed a new state oil company, Petoro. The goal of the company was to increase statoil's competitiveness and to participate in offshore ventures in Africa and South America. In order to financially achieve its goals of expansion, Norway had to reorganize and privatize a portion of its national oil company.[13] On the opposite end of the spectrum from Norway are numerous developing countries who seek privatization as a way of cleaning up the political quagmire.

For too many of the state-owned companies, corruption within the company and the central government has become a major problem. The governments used funds from the state-owned oil company for personal expenses and to maintain power. By privatizing, anticorruption leaders could begin to eliminate the intimate connection between the country's largest company and the central government. Accusations of corruption became widespread, and stories about oil wealth lining the pockets of politicians became commonplace in national newspapers. For example, Mexico's national oil company has been accused of corruption repeatedly. The *New York Times* reported that the company lost at least $1 billion per year to corruption. Since President Vicente Fox took office in 2000, he has pushed for privatization as a way of routing out the corruption.

Privatization, however, does not guarantee the elimination or even the reduction of corruption. Take the case of Russia, where corruption regarding its oil industry made major headlines in 2004. Russia's second largest oil company, Yukos, has faced criticism from the national government. The company's chairman, Mikhail Khodorkovsky, has faced legal charges for tax evasion and fraud and may face up to ten years in prison. Some believe that Khodorkovsky established a legitimate business in 1997 and that his case is a politically motivated trial orchestrated by President Vladimir Putin, who has already determined the ruling. Russia's government states that Khodorkovsky has a history of illegal activity. During the 1998 financial crisis, he allegedly protected himself at the expense of his bank's members through the Russian mafia.[14] While some of his actions have generated criticism from Yukos's stockholders, people have also accused Putin of trying to get even with Khodorkovsky for funding an opposition party. In December 2004, the Russian government forced Yukos to sell off pieces of the company to cover back-taxes. The impact of this trial on Russia's industry has been detrimental.

As a consequence of Yukos's case, Russia's industry has suffered from a decline in oil production and investor confidence. Since the beginning of the scandal in the fall of 2003, when the government seized almost one-half of the company's assets, Yukos's production has shifted downward. Also, the accusations prevented a $15 billion takeover by Yukos of another Russian oil company, Sibneft (Siberian Oil Company), which was scheduled to take place in the spring. Yukos and Sibneft agreed in April 2003, but Sibneft called off the merger in December of that year.[15] If the merger had been successful, the resulting company would have become the fourth largest private oil producer in the world. These political entanglements and suspected acts of corruption negatively impact the success of a country such as Russia and squelch its attempts at economic gain from its most valuable resource.

At times it seems that developing countries create their own roadblocks to development, often by passing strict, uncompromising laws that at the time appear to be protective measures. The founding fathers of the state-owned oil companies wanted to make sure that the exploitative relationship they faced with the foreign oil companies before nationalization never again occurred. In the case of Mexico, the constitution explicitly bans the privatization of the national oil company. Mexican law also gave PEMEX exclusive rights to the exploration and production of Mexico's natural gas. In order to develop its natural gas industry, Mexico had to work around this legal roadblock. The passage of the 1995 Natural Gas Law allows private companies to participate in the downstream sector of Mexico's natural gas industry. The law, at the same time, prevents one company from controlling more than one aspect of the industry. President Fox created a contract between PEMEX and private investors known as the multiple service contract, through which companies worked with PEMEX in the development of the natural gas industry but did not take ownership of the resources. The situation in Mexico is not unique, as more and more state-owned oil companies are moving toward privatization.

As privatization spread across oil-producing countries, growing concern over the future of these industries has emerged. Analysts are asking whether the privatization trend of the 1990s is the equivalent of the nationalization trend of the 1970s in that there are major flaws that are not apparent at this time. Will oil-producing countries have more success with privatization than nationalization and, if so, why? Furthermore, have oil-producing countries really changed the balance of power within the global oil industry? If the height of these countries' control lasted only two to three decades, then perhaps the OPEC revolution did not have the long-lasting effect its members desired. In other words, can the level of power oil-rich countries hold in the world be regressing to pre-OPEC times?

Oil-rich countries are concerned that by accepting the structural adjustment programs, they will return to the same problems they had with the foreign oil companies, such as negative cultural invasion and political manipulation. They fear that the flow of oil profits will once again leave the country and go to the West. Also, environmental organizations such as Venezuela's Orinoco Oilwatch are against the privatization of PDVSA and its subsidiaries because it will reduce the group's ability to pressure for change. A foreign multinational oil company may not be easily swayed by the demands of the Venezuelan community. In fact, evidence within the past decade suggests that this is exactly the case (see Chapter 6). The issue of privatization is just one component of the larger question regarding the tenuous relationship between oil and economic development.

DOES OIL ENCOURAGE DEVELOPMENT?

Within the discussion of economic development, the fundamental conclusion that oil, regardless of location, history, or culture, has represented a mixed blessing for all oil-producing nations is inevitable. In many ways, oil has encouraged economic instability. Oil-rich nations seem to be stuck in an endless cycle of the government propping up one part of the system only to have another collapse. One step forward with the high prices of oil during the second oil shock was followed by two steps backward during the third oil shock. During the 1970s and 1980s, many oil-producing countries became more dependent on their oil revenue than ever before, and their populations became increasingly impoverished. In the eyes of the public, their government did little to assist them, so they took their financial fate into their own hands. Starting the late 1980s, a rise of oil smuggling and shipping emerged in oil-rich developing countries. In Nigeria, thousands of barrels per day have been illegally shipped out of the country as Nigerians increasingly seek illegal methods of making money. Illegal fuel siphoning from pipelines became a major problem as well, and was accompanied by the emergence of a thriving black market for petroleum products.

Realizing the severity of the situation, the national governments turned to the international community for help. When they sought advice and assistance from the World Bank and the IMF, their situations did not necessarily improve. Many oil-rich developing countries came to the conclusion that foreign aid simply does not work. But is this an accurate assessment? Indeed, foreign aid may have caused more harm than good to an already fragile economy because, like oil wealth, the borrowed money often only encourages frivolous spending and corruption among autocratic leaders. This, of course, is not the fault of the lenders. Until a solution is found, the relationship between oil and economic development will remain a tumultuous one.

FURTHER READING

A tremendous amount of literature exists on the topic of economic development, especially works that focus on developing countries. As a starting point, Part 3 of Ken Cole's *Economy-Environment-Development-Knowledge* (New York: Routledge, 1999) provides an excellent overview of theories pertaining to development. See also Stuart Lynn's *Economic Development: Theory and Practice for a Divided World* (Upper Saddle River, NJ: Prentice Hall, 2003). For more information about the World Bank, see its Web site (www.worldbank.org). For more information about the IMF, see its Web site (www.imf.org). While the IMF and the World Bank have helped many developing countries, they have also garnered a great deal of criticism. For discussions on how the IMF's policies hurt developing countries, see Robert Weissman's "Why We Protest," *Washington Post*, September 10, 2001, 21, and Manuel Pastor's "Latin America, the Debt Crisis, and the International Monetary Fund," *Latin American Perspectives* 16, no. 1 (Winter 1989): 79–110. For praise and criticism of the World Bank's role in the economic development of oil-producing countries through aid, see Cheryl Payer's *The World Bank: A Critical Analysis* (New York: Monthly Review Press, 1982). For more information about the OPEC Fund, see its Web site (www.opecfund.org).

For comparison of how the oil-producing countries used or misused their oil revenues during the 1970s, see Ragaei El Mallakh's *Petroleum and Economic Development: The Cases of Mexico and Norway* (Lexington, MA: Lexington Books, 1984), Alan Gelb and Associates' *Oil Windfalls: Blessing or Curse?* (Oxford: Oxford University Press, 1988), and Jahangir Amuzegar's *Managing the Oil Wealth: OPEC's Windfalls and Pitfalls* (New York: I. B. Tauris, 2001). Works such John Bacher's *Petrotyranny* (Toronto: Dunburn Press, 2000) highlight the connection between oil, war, and dictatorship. For a discussion on oil price subsidies in Nigeria, see Nereus I. Nwosu's "The Politics of Oil Subsidy in Nigeria," *Africa* (Italy) (1996): 80–94, and on privatization in Latin America, see Ravi Ramamurti's "The Impact of Privatization on the Latin American Debt Problem," *Journal of Interamerican Studies and World Affairs* 34, no. 2 (Summer 1992): 93–125.

Chapter 9

Oil Politics and Its Global Impact

OPEC is held together by the idea that oil is going to run out. But the fields are refilling, one reservoir under laid by another for a long way down.

Thomas Gold, 2000[1]

Petroleum represents one of the most diverse and multi-functional minerals in the world. It has played an integral role in shaping global events, technology, and society for more than 2,000 years. Today oil intimately links the world together politically and economically, for better or for worse. While most people see petroleum as our primary source of fuel, its derivatives are equally important because they appear in almost everything we use in our daily activities. Among the wide range of petroleum by-products, people rely heavily on products made from rubber or plastic range from household utensils to clothing items. In short, it has become an essential part of our everyday lives. Given the importance of oil, it is not surprising that the world does not take the industry lightly and it has become a highly politicized commodity. In short, oil has become an essential part of our everyday lives. Given the importance of oil, it is not surprising that the world does not take the industry lightly; oil has become a highly politicized commodity.

Throughout the book, we have introduced several points that connect oil and international politics. Beginning with the discovery of commercial quantities of oil in the late 1880s, we saw the emergence of the major oil companies, which still dominate the industry today. These oil companies expanded their operations during the first half of the twentieth century throughout the world in search of new sources of oil and natural gas. New to the oil companies, oil-rich countries did not always take the necessary steps to maintain control over their valuable resource. As a result, tension between host nations and major oil companies developed, creating a long-standing struggle between the two, primarily from the 1950s to the 1970s. It was during this turbulent time that oil-producing nations took control over their industries as well as the global oil market through the creation of international alliances and state-owned oil companies. While this solved the problem of national control, several problems remained, and new ones emerged over time.

Because oil represents a mixed blessing for all oil-producing nations—regardless of location, history, and culture—these countries have had to deal with an array of problems. Nationalizing their industries did not remove the problems of environmental destruction, human rights violations, or economic instability. In fact, these issues became increasingly more difficult to resolve. The bulk of these countries lacked the necessary capital and infrastructure to maintain economic growth or expand the oil industry. In many cases, oil discoveries and developments occurred amid political instability and territorial uncertainty. With the exception of a few cases such as Norway, the major oil-producing countries are considered developing countries, and developing countries often share the common problem of ethnic conflict and poor social services. In many cases, civil societies in these countries suffer from poverty and government repression. For this reason, environmental and human protection have not been high priorities. Examples of community displacement, strikes, and sabotage related to the oil industry sprang up everywhere. This economic and political instability spread across borders and led countries into conflict over oil. The desire for control over oil fields even led countries to war, which only exacerbated the political and economic problems these countries faced. Whether these events occur nationally or regionally, they affect consumers all over the world.

Today hardly any part of the world is immune to the interplay of oil and international politics. Consumers feel the impact of national and international events that increase or restrict the supply of petroleum. Even if supplies do not actually change, prices reflect consumers' and producers' feelings of security about the oil industry. The threat of war within a major oil-producing region can increase the price of oil, as consumers and producers feel uneasy. While we have already covered many of the factors that affect the price of crude oil, and ultimately the price at the pump, we will introduce two more: levels of consumption and production in the present and future.

The discussions of consumption and production are nearly inseparable because one affects the other. In turn, both greatly affect the price of petroleum to the consumer. These two factors are crucial to the existence of the entire international petroleum industry. Without either of them, the importance of oil ceases to be. More important, these factors determine the future of oil and its price. If the global demand for oil decreases, then the price will adjust downward accordingly. Conversely, if demand increases, then the price will move upward. When demand outpaces production or production exceeds demand, the price of oil also adjusts to reflect these changes. For this reason, oil companies, stock traders, and OPEC all watch the levels of world consumption and production closely. Each attempts to predict the world's demand and supply to forecast trends in the world market.

WORLD CONSUMPTION

The single greatest determinant of world oil consumption is the world's current and future population. The world population in 2004 was just over 6 billion and is expected to increase by around 1.3 percent per year. With this projection, the global population by 2050 will reach 9 billion.[2] An increase in population directly affects the level of oil consumption because human demand for key materials such as petroleum and petroleum by-products will grow in proportion to the population. But population increase accounts for only a portion of demand.

The second greatest determinant of consumption is the economic growth and development of a country. On average, the standard of living for the majority of the population has increased. Over the years, industrialization has spread to more and more areas of the world. Also, the expanse of personal wealth in the world has increased. As a result, a greater percentage of the population consumes energy derived from petroleum as well as goods made from petroleum by-products. In the past, the typical family in the United States shared one automobile; today, almost every family member of age drives his or her own. And the recent trend in automobile size suggests that consumers in the United States believe that bigger is better. Automobiles in the United States appear to be increasing in size and decreasing in fuel economy. The trend of SUVs that began in the 1990s has only gained momentum in the twenty-first century, making them the top-selling vehicle for the United States. It is no longer unusual to see the once military standard vehicle, the Hummer, on city roads

Figure 9.1
World Consumption of Petroleum Products, 2003

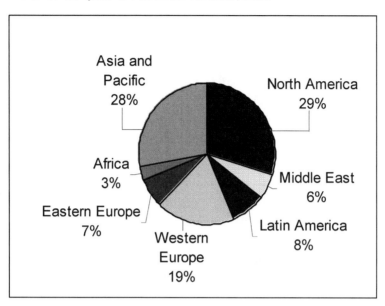

Asia and Pacific 28%

North America 29%

Africa 3%

Middle East 6%

Eastern Europe 7%

Latin America 8%

Western Europe 19%

Source: OPEC Annual Statistical Bulletin 2003 (Vienna: OPEC, 2004), 22. http://www.opec.org.

driven by the general public. The average SUV burns fuel at the rate of 15–20 miles per gallon of unleaded fuel (some hybrid cars currently on the market average 40 or more miles per gallon).[3] As a consequence of this overall new standard of living, the level of global consumption of petroleum products has increased.

Total world oil consumption was 81.7 million bpd in 2004 and is expected to increase by 1.9 percent per year (see Figure 9.1). According to the EIA's *International Energy Outlook*, global consumption will reach nearly 121 million bpd in 2025. A few regions of the world are attributed with generating the high demand for petroleum products. Rapid industrialization coupled with skyrocketing population growth has placed China and India among the group of high consumers, which includes the United States and Western Europe. In fact, China is expected to become the second largest consumer after the United States within the next couple of years.[4]

Almost since the discovery of oil, the United States has taken first place as the largest oil-consuming country in the world. According to the EIA, demand in the United States is projected to increase at a rate of 1.5 percent per year. Based on this projection, by the year 2025, U.S. consumption will reach almost 30 million bpd. In 2004, however, consumption increased by 3.5 percent.[5] Based on these projections, the United States will hold its position as the largest consumer for years to come.

Aside from population growth, the increase in demand can be attributed to a lack of energy-saving policies. Conversely, demand in Western Europe is expected to grow at the slower pace of 0.3 percent per year due to low population growth and the increasing cost of transportation fuel through taxation. With the projected overall increase in global consumption, oil companies and oil-producing countries have found it imperative to find new sources of crude oil and to improve the use of natural gas a source of energy.

NATURAL GAS PRODUCTION

Natural gas production and usage have become the latest trend in energy. While natural gas's potential has been known for quite some time, it was not until the 1990s that a major interest developed among producers and consumers. Today, it is expected to be the fastest-growing source of energy in the world, especially in industrialized countries. It is currently consumed at a rate of roughly 100 Tcf per year. World consumption is expected to increase by an average of 2.2 percent annually. By the year 2025, natural gas consumption is expected to reach 151 Tcf per year. This jump represents an almost 70 percent increase since 2001.[6] The rise in its popularity comes not only from a concern over the future of crude oil availability, but also from an overall increase in environmental consciousness. By using natural gas, consumers emit fewer greenhouse gases and other harmful pollutants into the air. While there are many convincing reasons for the use of natural gas over crude oil, the development of this industry is only beginning and has many obstacles to overcome.

The development of natural gas, first and foremost, can really only take off when demand and supply are relatively equal. As it stands, only about 20 percent of the world's known natural gas reserves are being exploited. It was estimated in 2004 that the world has proven reserves totaling roughly 6,000 Tcf. An exact number is difficult to obtain because no one will know exactly how much natural gas exists until it is extracted. Nonetheless, producers have come up with estimates. Using approximate numbers, experts project that the world's largest natural gas reserves are in Russia, at 1,680 Tcf. Saudi Arabia holds the fourth largest reserves at

225 Tcf, the United States stands in sixth place with 187 Tcf, and Nigeria at seventh with 159 Tcf. In short, the projected consumption is much higher than current production levels. To meet global demand, natural gas production must undergo major development and transportation projects worldwide. This, unfortunately, is where the majority of obstacles exist.

For the most part, oil-producing countries are the natural-gas-producing countries of the future. They are the same countries that struggled to develop a crude oil industry. Natural gas production has the same problems as crude oil production: location of oil fields in relation to markets, transportation, availability of investment capital, and technological expertise. In the early period of crude oil production, the associated natural gas that leaves the ground with the crude oil was flared off into the air. Even today, with the widely accepted advantages of trapping natural gas, around 10 percent of associated gas is still flared. Countries will continue to do this until marketing prospects for the resource improve.

What makes the development of natural gas particularly difficult is that the natural-gas-producing countries are doing what the foreign oil companies did during the twentieth century—creating markets for their products. For crude oil, the foreign oil companies accomplished this. When the major oil-producing countries gained control of their oil industry, many of the markets were already in place. In essence, the countries needed to focus only on production, refining, and distribution. With natural gas, they must do those three tasks as well as encourage the usage of the product globally and domestically. As any good marketing company knows, the appeal of natural gas will have to be low cost, environmental friendliness, and convenience. The bottom line, however, is that while demand for natural gas is increasing, it is not anywhere near that for crude oil. Until cars throughout the world run on compressed natural gas, a reduction in global dependence on crude oil will continue to drag. Natural-gas-producing countries will progress at the same pace.

Global demand is not the only challenge oil- and gas-producing countries face. Natural gas production requires a great deal of investment and technological expertise. Oil-producing countries can stop flaring the natural gas and begin trapping it. But this requires investing in the equipment and storage space to do so. Natural gas is associated not only with oil fields; geologists have found entire fields where natural gas exists without crude oil. Natural gas fields, like crude oil fields, take a great deal of exploratory effort and exist only in specific places throughout the world. Almost the half the entire world's proven natural gas reserves are located in the former Soviet Union. The Middle East is home to about 25 percent, and North America (Mexico, the United States, and Canada) makes up less than 10 percent. One major obstacle to developing the natural gas industry is that all of these natural gas reserves are located in somewhat remotes parts of each country. In Mexico, the major natural gas fields are located offshore. Within the former Soviet Union, most of the natural gas fields are located north of the Arctic Circle in Russia's eastern Siberia. The arctic climate, difficult terrain, and lack of roads make production a challenge. When and if production takes place, oil companies have the additional challenge of getting the natural gas to distribution centers and markets.

Natural gas fields are often located far from sizable markets, which poses a problem for transporting the natural gas. Miles of pipeline and roads and fleets of tankers are required to transport the natural gas produced. As it stands now, transporting natural gas via tanker in liquefied form or via pipeline in gaseous form costs four or five times more than shipping crude oil. For now, these costs do not make production within these regions cost-effective. This is not to say that demand for natural gas will not develop as crude oil fields mature and associated natural gas no longer flows from them. But natural gas production requires a great deal of investment on the parts of oil companies and producing countries.

Unfortunately, the experience of developing a natural gas industry is similar to the challenges oil-producing countries and companies faced with crude oil production. In other words, the same problems of a lack of capital and technological expertise they faced before are haunting them again. Venezuela, for example, holds slightly less than 150 Tcf, more than half of South America's natural gas reserves, but lacks capital and technology to exploit and distribute it. As a result, Venezuela flares into the air and reinjects into its oil fields billions of cubic feet a year. As with crude oil, oil- and gas-producing countries know the economic potential of natural gas production but need to invite foreign oil companies to invest. Recently Mexico invited foreign oil companies to engage in a service contract to assist in the transportation, storage, and distribution of its natural gas. Aside from production and transportation costs, the refining process, like that of crude oil, requires highly advanced facilities and expertise.

Once extracted from the ground, natural gas can be processed into a liquefied or gaseous form. Liquefied natural gas (LNG) is the most commonly used form because it transports easily. It is formed by reducing natural gas to an extremely low temperature, $-260°F$ ($-160°C$). The process of creating LNG requires special facilities that can be attached to crude oil refineries. Several oil-producing countries are currently building LNG plants. For example, Norway has planned for the construction of an LNG facility on Melkoye Island as part of its Snøhvit project. Norway plans to process the liquefied gas and export it to Spain, France, and the United States. The project is expected to get under way in 2006. The most challenging aspect of developing a natural gas industry has been the transportation of the product. For now, natural gas is a regionally traded commodity heavily dependent on pipelines and local tankers. The price of natural gas, therefore, is not as uniform as that of crude oil. Over time this will change as natural-gas-producing countries expand their markets globally. Russia, for example, now exports its natural gas to Western Europe, making it one of the largest suppliers.

The world's natural gas industry is also vulnerable to depletion. Although it seems premature to discuss this at the present time, countries need to discuss this possibility. Much of the world's natural gas is derived from crude oil fields, which means that there is always a risk that if a crude oil field collapses it may make extracting the natural gas no longer economically feasible. Therefore, depletion of crude oil and natural gas are being discussed together. Natural gas is not entirely considered an alternative source of energy to crude oil. Instead, it is seen as an extension of crude oil's versatile, but exhaustible, role in society.

DEPLETION

The concern for depletion of the world's oil and natural gas reserves is not new. The first formal emergence of it came from a group called the Club of Rome in 1972. The club was composed of concerned experts who met to discuss the situation in Rome, Italy. From this meeting came its publication entitled *Limits of Growth*, which stated with scientific confirmation that oil and natural gas reserves were finite. Using the rate of consumption that was taking place at the time, they calculated that these resources would be exhausted in about thirty years. While their calculations did not take into account the discovery of new oil fields, they definitely drew international attention to the fact that the world had become heavily dependent on a nonrenewable resource.

The important question regarding depletion, naturally, is when it will happen. What makes the year of absolute depletion difficult to determine is how reserves are defined. In producing crude oil and natural gas, countries and companies discuss the size and value of the fields in terms of their *reserves*. This term means that a field (which may include several wells) is estimated to have an amount in barrels for crude oil or cubic feet for natural gas. The trouble with discussing reserves is that the method of measurement varies. Legally, companies (private or state-owned) cannot include reserves for which no money to drill has been allocated. Oil in the ground maybe too costly to extract and once extracted not valuable on the world market. This kind of field is not supposed to be included in reserve calculations.

Not all companies or countries, however, abide by this method of measurement and often inflate the amount of their reserves. As with most of the oil industry, politics are involved. Countries and companies have both over the years been accused of exaggerating the amount of their reserves. Countries feel compelled to falsify reserves so that they look economically more stable than they really are, since many oil-producing countries rely heavily on oil revenues. Companies exaggerate in order to make shareholders feel secure in their investment. For these reasons, determining the time of depletion is difficult.

Forecasts have been made by companies and oil-related organizations regarding the depletion of oil and natural gas. For the most part, they agree that at the current rapid rate of consumption, depletion will occur within the middle of the twenty-first century. Based on current proven reserves of the world's petroleum (1,200 billion bbl of crude oil and 5,500 Tcf of natural gas), we can expect that another thirty to forty years of supply will be available. Of the total reserves, OPEC holds 891,000 million bbl. Divided into regions, the world's crude oil reserves break down as follows: North America has 27,200 million bbl; Latin America, 116,437.5 million bbl; Western Europe, 18,400 million bbl; the Middle East, 735,900 million bbl; Africa, 105,500 million bbl; Eastern Europe, 88,300 million bbl; and Asia and the Pacific, 46,000 million bbl.[7] Many of the fields included in these numbers are maturing. In the realm of natural gas, the world's total proven gas reserves may increase as more fields are developed in the near future. For now, production and consumption are keeping pace, assuming no new political developments obstruct global production. If, of course, consumption shifts up or down, then the time to depletion will decrease or increase. New discoveries and improved technology that will allow the extraction of uncounted supplies of petroleum will also change the predicted year of depletion.

Currently, Saudi Arabia is the world's largest crude oil producer. Its fields, however, are now in decline. Today, Saudi Arabia produces about 10 million bpd, but in order to meet the projected level of consumption, in the near future it will have to increase its production to almost 14 million bpd. Saudi Arabia is not necessarily running out of oil, but its fields are maturing, which makes extraction more expensive. Also, no new fields are being discovered. The duration of Saudi Arabia's fields largely depends on the increase in demand. As the threat of depletion intensifies, the world looks to the United States, which consumes around one-quarter of the world's supply and, hence, has tremendous influence over the speed of depletion.

As discussed in Chapter 8, many of the oil-producing countries do not have a reliable source of revenue outside of the oil industry. During the oil boom, several countries experienced the collapse of their agricultural sector. At the same time, the population became hooked on the influx of ready-made imported items. Import substitution projects that would promote the development of comparable products within the country did not always reach fruition. Oil-producing countries find themselves in a dilemma. If they have only a set amount of crude oil and natural gas in the ground, without any certainty of discovering new reserves, then how much conserving should they do? These countries run the risk of depleting their reserves early and missing out on high prices when global depletion nears. On the other hand, if they keep their oil in the ground too long, they may be stuck with unmarketable resources if the world finds an alternative energy source.[8] In any case, countries need to adopt a strategy that cushions the impact of absolute depletion.

Oil-producing countries have begun to develop policies that will ensure a more secure economic future without oil. Thus far, Norway has represented the model country. Norway's oil is predicted to run out within the first half of the twenty-first century, which is slightly earlier than the prediction for global depletion. To prepare for this, Norway created a Petroleum Fund to lessen the impact when oil revenues decline. Beginning in 1990, Norway began paying annually into the fund. Each year Norway pays in roughly $30 billion. Thus far, Norway has accumulated around $131 billion, and predicts that it will reach almost $400 billion by 2010.[9] Other oil-producing countries have not moved as quickly as Norway in protecting themselves for the future. Unfortunately, depletion can be relieved by only one thing: finding more crude oil and natural gas fields.

FINITE OR INFINITE?

Since the discovery of oil, debates have emerged over the true origin of petroleum. Geologists for the most part believe it is a *biogenic* resource, meaning that petroleum is derived from the remains of living organisms. According to these scientists, hydrocarbons formed from the application of heat and pressure to decomposing organic material deep under the Earth's crust. Once "cooked," the oil and gas travel to the Earth's crust and settle in sedimentary rock formations. This is why petroleum geologists survey the world looking only in sedimentary basins for oil and natural gas. A select number of geologists, however, do not believe in the biogenic origin of petroleum.

Beginning in the 1950s, a group of geologists, particularly in Russia, argued that petroleum is an *abiogenic* resource, which means that petroleum developed from a chemical process that did not include organic origins. They argued that under high temperatures and pressure, hydrogen and carbon developed in the Earth's mantle during its formation. These hydrocarbon molecules gradually leaked up to the surface through cracks in the rocks. This explains why they are primarily found in sedimentary basins throughout the world. Because hydrocarbons are not biogenic, they are embedded all over the world. The theory proposes that petroleum is not only abundant but nearly inexhaustible. Petroleum's availability, then, depends solely on a company's ability to extract it. To support this theory, Russian engineers drilled and struck oil and natural gas in western Siberia. While Russian scientists have been espousing this theory for decades, it has recently caught on in the United States and Western Europe through the work of the late Thomas Gold, who was a retired astronomy professor at Cornell University. He argues that part of the reason the theory has taken so long to catch on in the West has to do with the oil companies' interest in maintaining a fear of shortage, which allows for high oil prices on the world market.[10]

Within recent years the debate has intensified because the origin of petroleum will determine the probability that more petroleum will be found before the current sources run dry. The abiogenic theory, if true, could affect estimates of how much oil remains in the Earth's crust. If the origin of petroleum is, in fact, abiogenic, hydrocarbon formation will be endless, providing consumers with a limitless supply. The quest for an alternative source of energy would be unnecessary. To environmentalists, however, the search for a more eco-friendly energy source would continue. The abiogenic origin theory of oil formation is, however, rejected by most geologists. Regardless, all of these issues—consumption, production levels, and depletion—occupy the minds of the companies and countries whose future depends on solving these issues. The consumers, of course, are much more concerned with the impact on fuel prices.

PRICES AND INTERNATIONAL POLITICS

While not every country produces oil, every country consumes it. As a consequence, petroleum consumers all over the world feel the effects of fluctuating prices. The price of petroleum is the ultimate indicator of how consumers, through the trading among stockholders, view the future stability of the oil industry and respond to international oil politics.

Petroleum pricing reaches our attention at two stages: the price of crude oil on the global market and the price of fuel at the pump. The price of crude oil on the global market is determined by the trading that occurs on the New York Mercantile Exchange (NYMEX) and the International Petroleum Exchange (IPE). Within these exchanges, petroleum products are traded in variety of ways. Of particular importance is the *futures contract market*, where traders buy petroleum for the future based on the current market price. The futures market is a fast-pace trading system, created to reduce the risk of buying and selling petroleum products. Because these contracts are based on the future and are settled everyday, they reveal the most about how stockholders view the future of the global petroleum industry.

In general, OPEC tries to maintain the price of crude oil between $22 and $28 per barrel by setting target levels and production quotas for its members. Crude oil prices ranged between $2.50 and $3 per barrel from 1948 through the end of the 1960s. From 1958 to 1970, prices were stable, at about $3 per barrel. During the removal of Saddam Hussein in Iraq, the price of crude oil averaged more than $30 per barrel on the NYMEX.

Petroleum prices at the pump reflect the price of crude oil on the market plus the cost of taxation, refining, and transport to a specified region. The price of retail gasoline includes taxation from the federal, state, and local levels. In the United States, for example, these taxes account for roughly 30 percent of the cost of a gallon of gasoline. Refining costs make up about 13 percent of the retail price, which includes the cost of transporting the crude oil supply to the refinery. A large part of the cost has to do with the location of the markets. For example, in the United States, markets located far from the gulf coast, where the majority of the oil is imported, tend to have higher prices. The price also includes shipping costs from the refinery to the market. Seasonality also influences the retail price. For example, in the United States prices usually rise before and during the summer because people tend to drive more due to the favorable weather and vacation time. Environmental standards can also affect the price of gasoline. In California, for example, state law requires the use of reformulated gasoline, which reduces the level of toxins released into the air. As a consequence, gasoline in that state costs several cents per gallon more than elsewhere. These factors explain why fuel prices vary from country to country, state to state, and city to city.

Many countries produce oil specifically for domestic consumption and sell it at discounted prices to their local consumers. For example, Nigeria sets aside petroleum for its domestic consumers, and reduces the cost of its petroleum products to make it affordable. By subsidizing the cost to the consumer, Nigeria also makes the price of petroleum products uniform throughout the country. The government uses the subsidies to cover the transportation costs of trucks hauling oil long distances to remote parts of the country. Nigeria's low prices are attractive not only to Nigerians, but also to citizens of the neighboring countries of Benin to the west and Niger to the north. The result is that a substantial amount of Nigeria's oil gets smuggled out of Nigeria across the border to sell. The same problem happens in reverse as well.

Gasoline prices in Mexico, for example, are relatively high for North America because of various factors, including taxation. As a result, Mexicans often cross to the United States or Belize to purchase their petroleum products at a lower price. In response, the Mexican government began a pilot program in May 2002 aimed at reducing this cash outflow by reducing prices at test locations along the country's northern and southern borders. While countries such as Mexico and Nigeria display some sense of control over petroleum prices within their borders, they are subject to fluctuations of price on the international market just like everyone else. The interplay of oil and international politics does not leave any country untouched.

But what happens when oil is taken out of the equation because of global depletion? Without oil, would the environment be spared and human displacement become a thing of the past? Would international disputes over the location of mutual borders subside? Finally, if countries can no longer produce crude oil and natural gas, will an alternative source of energy carry on the legacy of oil? Many of these questions seem to address problems far into the future. After all, natural gas production is on the rise and crude oil is not expected to run out for thirty or forty years, barring any new developments. The reality, however, is that these issues are arising now.

In 2004, a new phenomenon occurred in the realm of petroleum pricing. A high demand for oil has caused an increase in price. Debates are emerging over this problem as to whether this is, in fact, the fourth oil shock or not. If it is, then economists fear that a recession will follow in 2005. Some economists argue that this is not a shock in prices, but a long-term trend due to consumption. The price of a barrel of crude oil began creeping up during the removal of Saddam Hussein from power in 2003. Most people attributed the high prices to the squeeze on other producers in meeting global supply as well as a general uneasiness on the trading floor.

But since the handover of power to the interim prime minister of Iraq, Iyad Allawi, the prices have continued to lurch upward, setting record highs. In August 2004, a barrel of crude oil hit almost $50 and remained in this range until the middle of 2005. Futures contracts determined on the market floor indicate that the high prices are expected to stay; con-tracts priced for six years down the road indicate that the price of oil will continue to hover around $35 per barrel. This is a major divergence from the many years when the price of crude oil stayed between $20 and $25. The *Economist* reported that demand was so strong in 2004 and the abil-ity to meet it so uncertain that traders joked about there being a "risk pre-mium" of $10 per barrel added to the price of crude oil. Part of the problem, particularly in the United States, is that spare capacity is almost nonexistent. In other words, if a shortage occurs, the world may not have enough reserves to cope with the shortage.[11]

Despite the potential oil crunch that has been predicted to occur, oil companies have not responded by investing in new oil and gas produc-tion. The high prices of oil during 2004 flooded the oil companies with money. The *Wall Street Journal* reported that while their cash flow in-creased 28 percent, their spending inched up by less than 10 percent.[12] Out of fear of a possible oil glut, oil companies have resisted outside pressure to expand the world's oil production. Likewise, OPEC refused to adjust its members' production because the price increase does not harm them. The group also holds fast to its method of price control and is hesitant to im-plement major changes.

The situation in 2004 strongly resembles that of the 1970s just before the oil embargo that caused the first oil shock. The world holds a spare oil-pumping capacity of about 1 million bpd, just as it did on the eve of the 1973 oil embargo. The problem, of course, is that consumption today has risen dramatically since that time. Since the early 1980s, fewer oil rigs and refineries are in operation than before. Furthermore, the United States is highly dependent on oil imports, which have escalated to 55 percent of the country's total usage. As discussed in Chapter 1, new oil and natural gas projects do not happen overnight. It can takes months, if not years, before a new field or refinery is up and running. If prices remain this high, it may be economically feasible for oil companies to begin producing from fields otherwise not considered usable. Nonetheless, immediate plans for refinery expansion or producing new fields could not alleviate the current situation.

All of the symptoms of an oil shock appeared to be in place, including how oil-producing countries are spending the influx of oil revenue. Are oil-producing countries destined to repeat their spending activities of the oil shock years? It appeared so. As high revenues flowed into oil-producing countries, strikingly similar spending patterns emerged. For example, after President Chávez of Venezuela won a recall election in August 2004, he used Venezuela's petrodollars to invest in his biggest supporters, the poor, by spending $1.7 billion on projects such as housing subsidies for them. This time around, however, the impact could be worse. Today many of these oil-rich countries are more in debt and more reliant on oil revenues than ever before. Since the 1980s, countries such as Nigeria have shifted from 80 percent reliance on oil revenues to 90 percent. The situation in oil-producing countries, overall, is more unstable than it was during the 1970s.

This new development in 2004 has raised serious questions about the future of oil and how this will shape the relationship between oil and international politics as we enter the twenty-first century. Are we moving forward and evolving to a more stable oil environment, or are we doomed to repeat the same fundamental problems that make oil both a blessing and a curse?

FURTHER READING

For more on the impact of oil prices on the global market, see Francisco Parra's *Oil Politics* (New York: I. B. Tauris, 2004). For theories and solutions to the pending problem of a global shortage of oil, see "The End of an Oil Age," *Economist*, October 25, 2003, 11–12, and Kenneth S. Deffeyes' *Hubbert's Peak: The Impending World Oil Shortage* (Princeton, NJ: Princeton University Press, 2001). For information on the debate over the origin of petroleum, see an interview with Thomas Gold in Oliver Morton's "Fuel's Paradise" *Wired*, July 2000, 160–72. For an examination of Saudi Arabia's role as a swing producer, see "Still Holding Customers over a Barrel," *Economist*, October 25, 2003, 61–63.

Part III

COUNTRY CASE STUDIES

Chapter 10

Iraq

IRAQ
Official Name: Republic of Iraq
Location: On the Persian Gulf between Iran and Saudi Arabia
Capital: Baghdad
Major Port City: Basra
Independence from British Protection: October 3, 1932
Population: 26 million
National Currency: New Iraqi dinar (NID)
Official Language: Arabic
Government: Interim President Jalal Talabani
OPEC Member: Since 1960
Proven Crude Oil Reserves: 115 billion bbl
Proven Natural Gas Reserves: 110 Tcf
National Oil Company: In transition
Crude Oil Production: 2 million bpd
Natural Gas Production: 90 Bcf per year
Production Ranking: Fifteenth largest (was ranked seventh until 2003)
Export Ranking: UN sanctions prevented commercial export until 2004

Map 10.1
Iraq

Within the past decade the economic and political landscape of the Republic of Iraq has experienced dramatic changes. With the removal of Saddam Hussein and the transfer of power from an interim government, the future of Iraq is uncertain. It will be no easy task for the new Iraqi government to solve preexisting social and economic issues. The population suffers from severe ethnic and religious conflict as well as problems of unemployment and poverty. The transition from the U.S.- and UK-led governing body to Iraq's own has so far gone relatively smoothly. To understand how Iraq has arrived at this state, it is important to look at Iraq's landscape and history.

Iraq is situated at the top of the Arabian Peninsula and shares its borders with Saudi Arabia to the south, Syria and Jordan to the west, Turkey to the north, and Iran to the east. Iraq benefits from its coastline on the

Persian Gulf as well as the two major rivers—the Tigris and the Euphrates—that flow through it. The rivers run parallel from the northwest to the southeast corners of Iraq and empty into the Persian Gulf. Iraq's topography varies from desert along the Saudi Arabian border to mountains over 3,000 feet tall along the Iranian border. The land between the rivers, Mesopotamia, was long considered the cradle of ancient urban civilization because of the emergence of wealthy city-states during the fourth millennium B.C.E. This area offers the most fertile land in Iraq. Following the rivers southeast to the Persian Gulf, the terrain changes into marshland. While Iraq boasts a somewhat diversified landscape, the majority of the country is covered by plains.

The territory that makes up Iraq is largely an artificial creation from a province known until 1921 as Mesopotamia. The Ottoman Empire occupied the territory from 1638 until its decline in World War I, when European powers drew arbitrary lines dividing British territory from French territory. The British first took the Kurdish-speaking region of Mosul, and then expanded their claim to bring in the Arabic-speaking Basra and Baghdad regions. This amalgamation of two cultures formed today's Iraq. From its formation, Iraq suffered from ethnic and religious divisions. The country's population consists of a Sunni Muslim Kurdish-speaking group in the north, a Shia Muslim Arabic-speaking group in the south, and a Sunni Muslim Arabic-speaking majority in the center. Iraq's official language is Arabic. The latter group dominated Iraq's government and received preferential treatment from the British during their rule.

After Britain claimed Iraq, it established an Iraqi monarch, who ruled the country under British control through a mandate system until the 1930s. Gradually, Iraqis, through a series of treaties, negotiated down the level of British occupation. Iraq did not have the unity or the money to challenge the British and declare independence, and so the kingdom negotiated its way out of British control. Iraq became a member of the League of Nations in 1932, which signified a giant step toward autonomy. But this semiliberation did not last.

World War II prompted another brief period of British occupation. The British relied on Iraq for oil and its strategic location, and needed political stability within the country to succeed during the war. For this reason, the British indirectly controlled almost all aspects of Iraq—from the military to its education system. Iraqis strongly resented the British presence during World War II, and nationalism escalated after the war. Iraqi nationalists criticized the country's monarchy for aligning itself with the

British. In 1958, the monarchy in Iraq was overthrown by nationalists, thus creating a new independent Republic of Iraq. This set the stage not only for independence, but also for a rash of military coups. In 1963 the Arab Socialist Ba'th Party (ASBP) ruled Iraq and remained in power until recently. One of its most prominent members, Saddam Hussein, came to power in the late 1970s.

Iraq under Saddam Hussein

Saddam Hussein came to power in 1979 and ruled until March 2003. Saddam Hussein al-Takriti, born in 1937, was an Arab Sunni from a poor family in northern Iraq. In the late 1960s he became a major player in the ASBP. He worked his way through Iraq's political system, holding increasingly important positions such as secretary-general and then vice president until 1979, when he took over as president. At the time, political power in Iraq was highly militarized, and the minority groups (Kurds and Shias) demanded liberation from Iraqi rule. Treatment of the Kurdish and Shia populations, coupled with Iraq's political arrogance, forced a war with Iran.

In 1980, escalating tensions between Iraq and Iran led to the Iran-Iraq War, which lasted for eight years and ended without a clear win on either side. While the war broke out due to a territorial dispute over who controlled the Shatt-el-Arab waterway, underlying tension had brewed for years. Competition for regional dominance as well as political and religious differences set the two countries at odds. The war began in 1980 when Iraq invaded Iran. Severely weakened by the revolution, Iran appeared an easy target for Iraq. Saddam Hussein envisioned the invasion to be brief and successful, but, instead, it lasted until 1988. The war ended in a stalemate, leaving both sides poorer and weaker than before the war began. Iraq's post-war state, however, did not prevent Saddam Hussein from launching another invasion.

In 1990 Saddam Hussein launched an invasion of Kuwait to seize control of its rich oil fields, beginning the Gulf War. Swift military action by the United States and its allies drove the Iraqis out of Kuwait in 1991. The United States sent over 500,000 troops into Saudi Arabia during the Desert Shield operation. In response to Iraq's aggression toward Kuwait, the United Nations decided to impose economic sanctions on Iraq and established "no-fly zones" in northern Iraq to protect Iraqi Kurds and Shi'ite Muslims in the south. The United Nations also placed pressure on Saddam Hussein to stop his persecution of the Kurds in the north and the

Shi'ite minority in the southeast. Although the Kurds relentlessly clamored for autonomy, Saddam Hussein had no intention of granting them independence because the Kurds lived near Kirkuk, one of Iraq's largest oil fields. The impact, however, hurt the Iraqi people and had little effect on the source of the problem. Ignoring pressure from the United Nations, Iraq challenged the protected zones and a U.S.-led alliance responded with bombs. The U.S.-led presence in Iraq continued through the rest of the 1990s to protect Kuwait and the oppressed groups in Iraq. In the late 1990s, a struggle emerged between Iraq and the United Nations regarding Iraq's refusal to disarm and allow weapons inspectors into the country. The more Iraq refused to allow weapons inspectors to survey Iraq's chemical facilities, the more frustrated the United Nations became with Saddam Hussein. Also, the United States believed that Saddam Hussein had a connection to the Muslim extremist group, al-Qaeda, which was the suspected orchestrates of the attacks in the United States on September 11, 2001. Because he was viewed as a threat to international security, a U.S.-led move toward his removal began.

The United States, working closely with the United Nations, concluded that Saddam Hussein's regime developed and continued to develop weapons of mass destruction. This concern prompted the UNSC to pass Resolution 1441 in November 2002. The resolution demanded that Iraq give up all its weapons of mass destruction or face "serious consequences." In February 2003 the United Nations still had not disarmed Iraq, and so the United Kingdom and the United States began pushing for military action. A month later, the United States and the United Kingdom launched a successful campaign to remove Saddam Hussein from power. After standing his ground until the last possible moment in Baghdad, Saddam Hussein fled and the country fell out of his control in April 2003. His whereabouts remained a mystery until December 2003, when U.S. forces found him hiding in a small space underground.

Post–Saddam Hussein Iraq

As of publication, Iraq has a new, civilian government for the first time since Saddam Hussein came to power in 1979. Between April 2003 and June 2004, the country was ruled under an interim government, known as the Iraqi Governing Council, under the Coalition Provisional Authority (CPA) led by the United States and the United Kingdom. Its main purpose was to oversee the creation of a congress and prepare for national elections. In June 2004 the official handover occurred making Iyad Allawi the prime minister of Iraq's interim government. In January 2005, an estimated 8 million Iraqis voted in a democratic election for a Transitional National Assembly. The outcome of the election gave a majority of the assembly seats to Shi'ites. Three months later the parliament selected the Kurdish leader, Jalal Talabani as president.

Most Iraqis are uncertain about the future of their country. Members of the various ethnic and religious groups hold differing ideas about what the new Iraq should look like. One thing that Iraqis appear to agree on is that they want the withdrawal of the United States from their country as soon as possible. They are eager to rebuild the country and get its oil industry back to full capacity on their own.

IRAQ'S OIL INDUSTRY

Discovery

Iraq's oil industry began with an Armenian entrepreneur, Caloust Gulbenkian, who formed the Turkish Petroleum Company (TPC) in 1914 to explore northern Iraq. The TPC was a consortium of Gulbenkian, Deutsche Bank, British Petroleum, and Anglo Saxon Petroleum Company (now a subsidiary of Royal Dutch/Shell). In 1925 the company received a seventy-five-year concession from the Iraqi government to operate throughout Iraq, which essentially was a British puppet under the mandate system. The concession agreement stated that TPC would remain under primarily British control and would grant the Iraqis royalties in exchange for oil. After drilling in northern Iraq, the TPC discovered in 1927 commercial quantities of oil in one of Iraq's largest oil fields, named Kirkuk, which formed the basis of Iraq's oil production.

In 1928 TPC and the Near East Development Corporation (a consortium of U.S. oil companies: SONJ, SONY, Atlantic Refining Company, Gulf Oil Corporation, and Pan American Petroleum and Transport Company) agreed to act as one company. To reflect this new company, TPC changed its name to the Iraqi Petroleum Company (IPC) in 1929 and invited two of its affiliates—Mosul Petroleum Company and Basra Petroleum Company—to secure additional concessions. Mosul was awarded a seventy-five-year concession in 1932 and Basra one in 1938. After the formation of the IPC, however, the company did not strike a large field until 1949, when it discovered two major fields, Zubair and Nahr Umr, in the southern tip of Iraq. From the mid-1950s through the 1980s, Iraq's oil production increased dramatically with the discovery of several new fields spattered across the eastern side of the country. Fearing nationalization, the IPC agreed in 1952 to a 50 percent profit-sharing arrangement, which allowed the company to stay in the country until 1961.

The Industry Today

Iraq's industry today is in fragile state because the country lacks the capital, the security, and the infrastructure to conduct repairs and run the day-to-day operations that are needed. During Saddam Hussein's rule, the oil industry did not run at full capacity because many of the facilities had been shut down or damaged by war, or functioned poorly. The oil reservoirs were poorly maintained and much of the equipment used was corroded from extended use. As a result, Iraq's production levels fluctuated from year to year, war to war. In the 1990 Iraq's production peaked during Saddam Hussein's rule at 4 million bpd, while after his removal in 2003 Iraq produced roughly 1–2 million bpd. Just as the political system in Iraq is experiencing a major overhaul, the oil industry is as well.

Under the new civilian government, Iraq's oil industry is expected to run at full capacity soon. Iraq's industry holds promise because it is said to have the lowest production costs in the world, which should attract oil investors. Iraq does not boast the highest quality of oil, but it offers an abundance of medium to heavy crude with sulfur. Consumption of oil is low in Iraq, making almost of its production available for export. And Iraq is believed to have sizable reserves, totaling 115 billion bbl of oil, and blocks of untouched fields along the western border.

In 1961 Iraq passed Public Law 80, whereby Iraq expropriated 95 percent of the IPC's concession area without compensation. Having gained public oil fields, Iraq announced in 1964 its decision to form the Iraqi National Oil Company (INOC) to take complete control of its oil industry. In 1972, the transition of control was completed. A year later the Iraqis agreed to settle outstanding claims with the IPC by allowing the company to retain concessionary rights in southern Iraq through a sister company, Basra Petroleum Company. Although this allowed IPC to remain, the company had lost Kirkuk, the largest oil field in Iraq. Within the INOC, the Northern Oil Company and the Southern Oil Company managed production operations in their respective parts of the country. Since the removal of Saddam Hussein, the future of Iraq's national oil company has been under consideration.

Just before Saddam Hussein's removal in 2003, several contracts had been signed or were in the negotiation process. The majority of the contracts were with French, Canadian, and Russian companies. The CPA decided to review these contracts and determine which are valid. It is not just the contracts, however, that stand in the way of Iraq's oil industry. Political instability in the country made oil pipelines and other facilities a target for sabotage and wartime destruction.

Iraq has several oil pipelines, of which three cross into neighboring countries, but none of the transborder lines are in operation. The Iraq-Syria pipeline (closed in 1982) and the Iraq–Saudi Arabia pipeline no longer function. In the 1930s, a pipeline was laid connecting oil fields in northern Iraq to Haifa, Israel, but Iraq closed it in protest of the establishment of Israel as a state in 1948. The Iraq-Turkey (also known as the Kirkuk-Ceyhan) pipeline runs from the Kirkuk oil fields to Ceyhan, Turkey. The pipeline suffers from repeated sabotage within Iraq. In 2004, plans were made for a pipeline from Iraq across the Shatt al-Arab waterway to the port of Abadan in Iran. Within the country, the pipelines connecting fields to refineries and export terminals have not operated regularly due to war and acts of sabotage.

What oil does flow goes to four possible terminals: Basra, Khor al-Amaya, Umm Qasr, and Khor al-Zubair, which are all located on the Persian Gulf. The Basra terminal is Iraq's main export outlet in the Persian Gulf and was used for smuggling oil during the UN sanctions against Iraq. Khor al-Amaya is the second largest shipping center, but was severely damaged during the first Gulf War. The refineries have been subject to the same inconsistent pattern of operations as the pipelines and terminals. A major problem has been the lack of security to protect the pipelines.

At full capacity, Iraq has eight refineries, which are all located in eastern Iraq near production or consumption centers. The three largest refineries—Baiji, Basra, and Daura—were severally damaged during the Gulf War. Interestingly, Iraq produced more oil during Saddam Hussein's rule than it could consume or export. To remedy the problem, Iraq frequently reinjected the crude oil back into the ground for storage. Today however, gasoline for domestic consumption is in short supply.

Iraq has eighty discovered oil fields, but only fifteen have been developed. The fields currently being drilled are mainly in a region north of Baghdad and in the south near Kuwait and the Persian Gulf. The northern fields include Kirkuk, Bai Hassan, Jambur, Khabbaz, Ajil (formerly Saddam), and Ain Zalah Butmah-Safaia. The southern fields include North and South Rumaila, West Qurnah, Az Zubair, Misan/Buzurgen, Majnoon, Jabal Fauqi, Abu Ghurab, and Luhais. Iraq's largest fields are Majnoon and Qurnah in the south and East Baghdad just outside of the capital city, while its most highly productive fields are Rumaila and Kirkuk. During the recent war, oil wells were set on fire in the Rumaila fields. Fortunately for Iraq, the amount of destruction in relationship to the number of fields (developed and undeveloped) was small. In addition to crude oil, its fields also provide natural gas.

While Iraq flares off much of its natural gas, it does capture a small percentage for domestic consumption and export. Its natural gas reserves total 190 Tcf, which includes associated and non-associated gas. The main associated sources are the Ain Zalah, Butmah, Kirkuk, Bai Hassan, Rumaila, and Zubair fields. The non-associated natural gas fields include al-Anfal, Chemchamal, Jaria Pika, Khashm al-Ahmar, and Mansuriya. Of these fields, al-Anfal is the only field currently producing natural gas. Iraq is in the negotiation process with multinational companies for production rights. Iraq has plans to expand its natural gas industry by building an LNG facility as well as laying new gas pipelines.

FURTHER READING

Since the 1990s, the number of books devoted to the history and politics of Iraq have increased dramatically. While many focus on the Gulf War and Saddam Hussein, a number of them provide an excellent overview of Iraq's history and culture. These works, in a sense, are focused on providing background to recent events. See, for example, Toby Dodge's *The Inventing of Iraq: The Failure of Nation Building and a History Denied* (New York: Columbia University Press, 2003), Shams C. Inati's edited work *Iraq: Its History, People, and Politics* (Amherst, NY: Humanity Books, 2003), and Charles Tripp's *A History of Iraq* (Cambridge: Cambridge University Press, 2002). For an overview of the nationalization of the IPC, see Michael Brown's "The Nationalization of the Iraqi Petroleum Company," *International Journal of Middle Eastern Studies* 10, no. 1 (February 1979): 107–24. After the Gulf War, several works, such as Hamdi A. Hassan's *The Iraqi Invasion of Kuwait: Religion, Identity, and Otherness in the Analysis of War and Conflict* (London: Pluto Press, 1999), and Phyliss Bennis and Michel Moushabeck's *Beyond the Storm: A Gulf Crisis Reader* (Northampton, MA: Interlink, 1998), emerged that outline the possible causes and different perspectives regarding its origins. See also Stephen C. Pelletière's *The Iran-Iraq War* (Westport, CT: Greenwood Press, 1992). For an excellent introduction to Islam, see Thomas W. Lippman's *Understanding Islam: An Introduction to the Muslim World* (New York: New American Library, 1982).

Chapter 11

Mexico

MEXICO
Official Name: United Mexican States
Location: North America, south of the United States
Capital: Mexico City
Major Port Cities: Salina Cruz and Veracruz
Independence from Colonial Rule: September 16, 1810, from Spain
Population: 106 million
National Currency: New Mexican peso (MXN)
Official Language: Spanish
Government: President Vicente Fox (since 2000)
OPEC Member: No
Proven Crude Oil Reserves: 15 billion bbl
Proven Natural Gas Reserves: 15 Tcf
National Oil Company: Petróleos Mexicanos (PEMEX)
Crude Oil Production: 3.8 million bpd
Natural Gas Production: 1.3 Tcf per year
Production Ranking: Fifth largest
Export Ranking: Ninth largest

Map 11.1
Mexico

The United Mexican States (Mexico) is the third largest Latin American country, with thirty-one states and a federal district that includes Mexico City. Mexico forms a large tongue-like shape that largely connects North and South America and acts as a divider between the Pacific Ocean and the Gulf of Mexico. It shares a 1,900-mile border with the United States and a smaller border with Belize and Guatemala to the south. The landscape of the country varies from desert and mountains in the north close to the southwestern United States to tropical rainforest terrain in the south. Mexico is situated in a geologically unstable area, with parts of Mexico subject to volcanic eruptions and earthquakes. In 1985 an earthquake hit Mexico City, causing a great deal of damage. In addition to natural disasters, Mexico has also been plagued with economic and political problems. These issues, however, do not over shadow Mexico's rich history and culture.

Much of what makes Mexico unique is the country's diverse population. While over half of Mexico's population is considered *mestizo* (a mixture of European and native descent), about 30 percent is Amerindian or other ethnicities. Mexico's official language is Spanish, but about fifty Amerindian languages are also spoken. The majority of Mexicans live in or around Guadalajara, Mexico City, and Monterrey. Despite the government's efforts, a large number of Mexicans struggle to find work and adequate housing. The rural exodus has resulted in the formation of shanty towns and a decline in living conditions outlining Mexico's major cities. And many people living in Mexico are dependent on the money sent home from family members working in the United States. Mexico has little arable land and not enough industry to employ its population. As a consequence, large numbers of people migrate to urban centers in Mexico or to the United States in search of work. It is estimated that roughly 300,000 Mexicans emigrate to the United States (legally and illegally) each year.

Mexico's history largely falls into four major periods. The first period is the pre-Columbian—the period of Indian civilizations until Spanish conquest in the early sixteenth century. It represents a time of territorial expansion and conquest. Mexico experienced a succession of highly sophisticated civilizations such as the Olmecs (ca. 1300 to 400 B.C.E.), the Mayans (300 to 900 C.E.), and the Aztecs (1400 to 1521 C.E.), who built Tenochtitlán, but were destroyed by Spanish conquerors. The second period of colonial rule lasted until the nineteenth century, when Mexico begun its struggle for independence in 1810. Overthrowing Spain, however, took over a decade to complete making the year of absolute independence 1821. After independence, however, Mexico was not free from foreign occupation. In 1864 the French emperor Napoleon III established an empire in Mexico under the Austrian prince Maximilian of Habsburg until 1867. In 1867 Mexico defeated the French and established a new republic in 1872.

The third period includes the republican years, which ended in the bloodiest war in Mexico's history—the Mexican Revolution. Considered a peasant revolution, it began with a national uprising to oust President Porfirio Díaz. During years of fighting and social displacement, millions of people died as clashes took place in almost every city across Mexico. Although it began in 1910 as a political movement to overthrow a dictator, it turned into a social revolution. People clamored for land redistribution, protective labor laws, the nationalization of foreign companies, and the expansion of public health and education. Through the Constitution of 1917, many of these demands were met.[1]

After the overthrow of Porfirio Díaz, Mexico transitioned from "revolutionary" politics to "evolutionary" ones. The government's policies shifted from "radical socialism to industrial capitalism" in order to transform the country from agriculture to industry. The final period of Mexico's history spans from the 1920s to the present, during which Mexico has transformed rapidly into a modern nation grappling with issues of globalization, urbanization, and economic instability. The year 1940 marked the end of a military presidency in Mexico and the beginning of Mexico's involvement in global politics. During World War II, Mexico took an active role, supplying war materials such as petroleum to the Allied forces. The wave of nationalism that resonated through the developing world also hit Mexico, and state control of the economy and energy sector increased. In the 1960s, internal problems plaguing Mexico reached international attention. For example, in 1968 a student demonstration during the Olympics in Mexico City turned violent, killing and wounding hundreds of protestors. Also, the ruling party, the Institutional Revolutionary Party (Partido Revolucionario Institutional, or PRI, as it is commonly known) adopted a reputation as perpetuating a patronage system. While many had become wealthy through their involvement in the oil industry and national politics, the vast majority of Mexicans still lived in poverty. Industrialization efforts that had been financed through heavy government spending during the 1960s sent Mexico into serious financial trouble by the 1970s. When the oil boom hit Mexico, it encouraged the country's overreliance on its petroleum industry as well as corruption. The 1980s and 1990s represent Mexico's efforts to reign in and solve its economic problems through working with the IMF and implementing land reforms to encourage small farm ownership instead of cooperative farms. Most recently, Mexico has been rebounding from a recession caused by the devaluation of the peso in the 1990s and has undergone several changes to encourage foreign investment and domestic industry, particularly in the oil industry. Elected in 2000, Mexico's current president, Vicente Fox, has attempted to wipe out corruption and revitalize Mexico's oil industry.

MEXICO'S OIL INDUSTRY

Discovery

Early attempts at oil production in Mexico began when Emperor Maximilian awarded thirty-eight oil concessions to French and Mexican entrepreneurs in 1865. Several other unsuccessful attempts followed, including one by Cecil Rhodes of South African fame. Shortly after the discovery of oil in the United States, U.S. oilmen turned their attention to Mexico's potential. In the late 1880s, Henry Clary Pierce and William H. Waters brought their Standard Oil affiliate company, Waters-Pierce Oil Company, to Mexico. This company focused on building refineries and securing oil supplies for Standard Oil. Edward L. Doheny arrived in 1901 and formed his Mexican Petroleum Company and the Hausteca Petroleum Company. He went to Mexico under the direction of the Waters-Pierce Company but pursued his own oil interests and struck oil in the region of San Luis Potosí. While Doheny claimed no association with Standard Oil, he actually sold crude oil to the company. With the kind of success Pierce, Waters, and Doheny had, it was not long before they were joined by another oilmen. The Mexican government invited a British oilman by the name of Sir Weetman Pearson (who later became known as Lord Cowdray) to develop Mexico's railroad industry. After completing his requested work, he searched for oil and discovered in 1908 the famous "Golden Lane" fields. He formed his Compañía Mexicano de Petróleo el Aguila (referred to as simply El Aguila) to produce the fields. Pearson maintained a close relationship with Mexico's government and created a company that included members of the Mexican elite. Through this strategy, the company presented itself to the public as a national company. The four oilmen, Waters, Pierce, Pearson, and Doheny, dominated Mexico's oil industry, competing with one another until the major oil companies arrived.

Between 1910 and 1920 Mexico experienced a dramatic increase in its oil production as new oil companies arrived in the country yearly and discovered more oil fields. Among the many companies profiting from their ventures in Mexico was Royal Dutch/Shell. Prior to this so-called Golden Age of oil, Royal Dutch/Shell had expanded its presence in Mexico by purchasing Pearson's El Aguila in 1919 and SONJ, which purchased the independent British Compañía Petrolera de Transcontinental in 1917. Although Mexico's industry was dominated by Royal Dutch/Shell and SONJ, Gulf Oil and Texaco also staked their claims. By 1920, foreign companies owned or controlled virtually every aspect of the country's oil industry. Within a decade, Mexico had become the supplier of one-quarter of the world's oil. While Mexico's industry expanded, a national revolution was under way. The presence of the oil companies during this critical time created speculations regarding the political involvement. Animosity toward the oil companies escalated from all sides, particularly from the Mexican oil workers. Labor disputes with the major oil companies ensued and the government decided in 1938 to nationalize Mexico's oil industry.

Industry Today

Since the discovery of oil in Mexico, Mexico's oil industry has continued to expand. Today Mexico has the world's third largest proven oil reserves in the Western Hemisphere at 15 billion bbl. Mexico produces 3.8 million bpd of crude oil. Until the 1970s, Mexico was able to retain a sizable percentage of its daily output to meet domestic consumption without having to import oil. In 1971, Mexico could no longer satisfy both domestic and international demand. Mexico's entire oil industry is run and operated by the state-owned company PEMEX, which was created after nationalization in 1938. PEMEX conducts virtually all activities, from exploration to retail. It holds exclusive rights to oil exploration and production in Mexico on- and offshore, but within recent years has considered bringing in foreign companies on a service contract basis. In addition to production, it also operates pipelines, tankers, storage facilities, refineries, and retail outlets. Today, PEMEX is the world's eighth largest oil company.[2] Even after seventy years of operation, PEMEX is still an important symbol of national pride for Mexico.

Today Mexico produces a range of crude oil that varies from extra light to heavy, depending on the oil field. In general, Mexico's oil is considered heavy in comparison to others traded on the world market. Its oil trades as Isthmus, Maya, and Olmeca, just to name a few. The light crudes come from fields in southern Mexico, while the medium and heavy crudes are located in the southeast. Almost all of Mexico's oil comes from the "Golden Lanes," which run along the Bay of Campeche on the eastern side of the country. Fields north of the Golden Lanes date back to discoveries made in the 1940s and the southern fields include those discovered in the 1970s and 1980s.

Mexico's industry also includes offshore drilling, which began in 1963. These fields are collectively known as Cantarell (a complex of four fields) in the Bay of Campeche. This group produces about 1.9 million bbl of heavy crude oil per day. With the threat of decline looming, PEMEX continually explores for new fields. In 2003 PEMEX announced the discovery of fields in the Grijalva River Delta in Tabasco and Coatzacoalcos in Veracruz. Every oil field connects to one of Mexico's many refineries and export terminals.

Most of Mexico's refineries and export terminals are located on the eastern side of the country along the Bay of Campeche. The few exceptions to this include refineries in Mexico City and Salina Cruz, and several terminals that dot the west coast (Guayamas, Lázaro Cárdenas, and

Acapulco). On the west coast, the Mexican refineries and export terminals are located in almost every major city (Veracruz, Tampico, Minatitlán). Several of the refineries were built in the first half of the twentieth century and are undergoing renovations to modernize, and more important, increase their output levels. PEMEX has also made steps toward expanding its natural gas production.

Mexico has significant natural gas reserves of 15 Tcf, the sixth largest natural gas reserves in the world. Since the late 1980s, Mexico's consumption of natural gas has exceeded daily production, requiring the country to import it from the United States. Within the past few years, expansion of natural gas consumption and domestic production has become a major focus for the government. Mexico has used the development of natural gas as an experiment in privatization. The problem for Mexico has been a clause in the constitution stating that PEMEX has exclusive rights to the exploration and production of natural gas. In 1995 Mexico passed the Natural Gas Law, allowing private investors to be involved in the storage and distribution of natural gas, while still reserving ownership with PEMEX. In 2004, Mexico invited foreign companies into the country to expand its natural gas production. PEMEX offered these companies seven blocks in the Burgos Basin Province in the northeast. Companies such as Brazil's Petrobras received a service contract that will last for fifteen years. Mexico also produces gas from the southwestern fields of Chiapas and Tabasco. All of the natural gas fields are connected by a network of pipelines.

Mexico's natural gas pipeline network extends from southeastern Mexico to several points within the country and across the border to the United States. In total, Mexico has almost 6,000 miles of pipeline. Connections between the United States and Mexico via pipelines include Piedras Negras, Reynosa, Samalayuca, Ciudad Juarez, and two in Arguelles (all connecting to Texas), Mexicali (connecting to California), and Naco (connecting to Arizona). Within the past few years, Mexico developed a pipeline that connects Baja, Mexico, to California and has plans to erect several LNG facilities in cooperation with foreign oil companies. Overall, Mexico has taken a leading role in natural gas development among oil-producing countries through its decision to invite private oil companies instead of relying on PEMEX.

Within the past few years, allegations of corruption within the Mexican government and PEMEX have frequently emerged. The government has been accused of siphoning off oil revenue for personal use totaling at least $1 billion per year. Since President Fox came into office in December 2000, however, attempts to eliminate corruption within the system have been made.[3] The Mexican government is hopeful that with an expanding oil and gas industry and the reduction of corruption, Mexico's economic situation will improve.

FURTHER READING

For the history of Mexico, there are several good works available. For an overview of Mexico's history and culture, see Peter Standish and Steven M. Bell's *Culture and Customs of Mexico* (Westport, CT: Greenwood Press, 2004). For a brief history of Mexico, see Robert Miller's *Mexico: A History* (Norman: University of Oklahoma Press, 1985) and Burton Kirkwood's *History of Mexico* (Westport, CT: Greenwood Press, 2002). Within the eight-volume series *The Cambridge History of Latin America* (New York: Cambridge University Press, 1984–1995) are several good chapters that focus on Mexico, such as Freidrich Katz's "Mexico: Restored Republic and Porfiriato, 1867–1910," vol. 5, 3–78, and Jean Meyer's "Mexico: Revolution and Reconstruction in the 1920s," vol. 5, 155–96. For a close look at the Mexican Revolution, see John Womack Jr.'s chapter entitled "Mexican Revolution, 1910–1920," vol. 5, 79–153. For an excellent work on Mexico's oil workers, see Jonathan Brown's *Oil and Revolution in Mexico* (Berkeley: University of California Press, 1993). Several works have been written on the expropriation of Mexico's oil industry and the formation of PEMEX. See George Philip's *Oil and Politics in Latin America: Nationalist Movements and State Companies* (Cambridge: Cambridge University Press, 1982). For a look at PEMEX during the 1970s and 1980s, see Laura Randall's *The Political Economy of Mexican Oil* (New York: Praeger, 1989).

Chapter 12

Nigeria

NIGERIA
Official Name: Federal Republic of Nigeria
Location: Coast of West Africa between Benin and Cameroon
Capital: Abuja (since 1991)
Major Port Cities: Lagos and Port Harcourt
Independence from Colonial Rule: October 1, 1960, from United Kingdom
Population: 137 million
National Currency: Naira (₦)
Official Language: English
Government: President Olusegun Obasanjo (since 1999)
OPEC Member: Since 1971
Proven Crude Oil Reserves: 35 billion bbl
Proven Natural Gas Reserves: 176 Tcf
National Oil Company: Nigerian National Petroleum Corporation (NNPC)
Crude Oil Production: 2.5 million bpd
Natural Gas Production: 500 Bcf per year
Production Ranking: Thirteenth largest
Export Ranking: Eighth largest

Map 12.1
Nigeria

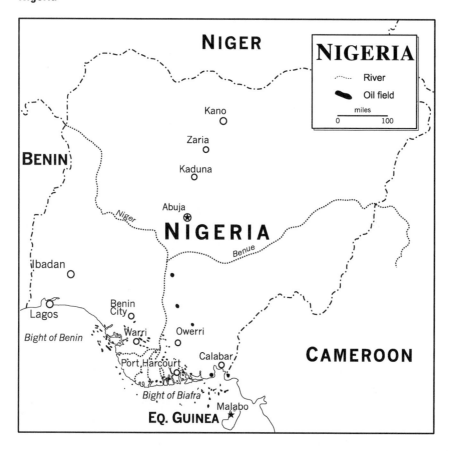

Nigeria is Africa's largest petroleum exporter and a major source of U.S. crude oil imports. Oil has had both a positive and negative impact on Nigeria. The industry generates a great deal of money and, if managed well, has the potential to turn Nigeria into a wealthy country. Despite the large amounts of oil being produced and the large amount of revenue collected from the oil industry each year, Nigeria continues to face economic hardships. The situation is so severe that Nigeria is considered one of the most corrupt and poverty-stricken countries in the world. Despite the political and economic instability, many regard Nigeria as a special place. It is the origin of many African slaves relocated to the United States during the trans-Atlantic slave trade. These slaves brought with them many cultural traditions, such as religion and cuisine, which continue to be an important part of U.S. culture.

The Republic of Nigeria is located on the west coast of Africa. Nigeria borders Benin to the west, Cameroon to the east, and Niger and Chad to the north. Nigeria is often described as the "giant" of the West African countries because of its size, military strength, and population. It is a multiethnic country, with more than 200 groups speaking numerous languages including its official language, English. The topography of Nigeria greatly varies from the North to the South. The northern region of the country is mainly grassy plains and semidesert because of its close proximity to the Sahara, while the southern region is lush with tropical vegetation.

The country known today as Nigeria came into existence as a formal political entity in 1914 under the authority of Lord Frederick Lugard on behalf of the British. Previously Nigeria had been divided into two British protectorates. As a colony, Nigeria had politically strategic and economic significance for the British. Its economy relied mainly on exporting Nigeria's natural resources such as peanuts and cocoa to Europe. The British sought to invest as little as possible in the colony, while extracting as much as possible from the land and its people.

To effectively rule the colony, the British used a policy of divide and rule, maintaining control through keeping the different ethnic groups apart. For this reason, the ethnic groups making up the Nigerian population did not develop into a strong nation during the colonial period. Nigeria was divided into three large regions that represented the three largest ethnic groups. Present-day Nigeria, though divided into thirty-six states to form a federation, still maintains this general layout. The northern region includes Nigeria's predominantly Muslim people, known as the Hausa-Fulani. The eastern region is where the majority of Nigeria's Igbo people live, and the western region is where the Yoruba live. Nigeria's history has been complicated by the ethnic rivalries that exist among the three regions.

Nigeria became independent on October 1, 1960. Only a few years after independence, severe political troubles began. The political history of Nigeria from 1960 until the present can be viewed as a continuous struggle between regional autonomy and central control. This ethnic-regional competition led to a continuous search for a constitutional formula to hold together the Nigerian Federation, and to an ongoing battle over the regional allocation of public revenue, most of which is derived from export goods such as oil. Between 1960 and 2000, Nigeria experienced seven military coup d'états, often accompanied by brutal military regimes, a civil war that lasted three years, and ongoing flare-ups of religious violence between Christians and Muslims in the North. These events were fueled by and continue to fuel ceaseless political corruption and economic instability. Nigeria has experienced roughly twenty-nine years of military rule and only fourteen years of democratic rule. During those short periods of democratic rule, accusations of fraud have been widespread, giving Nigeria the rank of second most corrupt country in the world.[1]

NIGERIA'S OIL INDUSTRY

Discovery

The year 1908 marks the first attempt to find oil and develop an industry in Nigeria. A German company, the Nigerian Bitumen Company, began to explore Nigeria's coast in search of bitumen. At this time, the British claimed Nigeria as a colony. In 1937, Royal Dutch/Shell and British Petroleum began operations in Nigeria under an official ordinance that granted them the right to explore both onshore and offshore. They explored what is today's major production center—the Niger Delta in the eastern region of Nigeria, which is one of the world's largest wetlands. It is named after the delta of the Niger River that pushes out into the Atlantic Ocean where today offshore drilling takes place. To explore the region, Royal Dutch/Shell and British Petroleum formed a joint venture company called the Shell-BP Petroleum Development Company of Nigeria Limited in 1946. Competing companies such as Exxon did not attain access to Nigeria's oilfields until the 1960s.

Aside from these early attempts, active exploration for petroleum did not begin until the 1950s. On behalf of British interests, Shell and BP became particularly motivated after the Egyptians nationalized the Suez Canal in 1956. The British owned substantial shares in each company and sponsored Shell-BP's exploration in Nigeria. The British heavily depended on the canal to transport oil from the Middle East to Europe. When the Egyptians temporarily denied the British passage through the Suez Canal, the British felt compelled to expand their oil production elsewhere.

After many years of surveying the land and drilling test wells, with substantial investment, Shell-BP finally struck oil in 1953 but not yet in commercial quantity. In 1956 the company struck oil for the first time in commercial quantity at Oloibiri and Afam. With this discovery, Shell-BP opened its operational headquarters in Port Harcourt, the largest seaport in Nigeria. Here the company set up housing with water and electricity available as well as training schools for the oil workers. It also built storage facilities for the crude oil. Two years later, the company exported the first cargo of crude oil from Nigeria.[2]

In the midwestern region of Nigeria, close to the Niger River, crude oil was discovered in 1958 at Ugheli. However, commercial production did not begin there until the mid-1960s, leaving the eastern region as the sole provider of Nigeria's crude oil. This changed during Nigeria's civil war because the federal government could not depend on the crude oil from the eastern region. The war from 1967 to 1970 culminated from an ongoing conflict among competing regional powers and ethnic groups. The eastern region, predominantly Igbo, seceded from Nigeria to form the Republic of Biafra. The secession had an enormous impact on Nigeria because the eastern region took much of Nigeria's oil industry with it. By 1970, when the war ended and the eastern region returned to Nigeria, the midwestern region had become a major oil-producing state. Nigeria's turbulent start as an independent country after nearly one hundred years of British colonial rule has greatly impacted the success of its oil industry.

The Industry Today

Since its discovery, Nigeria's oil has been in high demand on the world market because of its quality and abundance as well as its close proximity to Europe. The oil is low in sulfur and light in consistency, making it high in quality. Nigeria's proximity to Europe and the United States also gives Nigeria a competitive edge over the Middle East for marketing its oil. Currently only about 20 percent of Nigeria's oil production is consumed locally. Nigeria also supplies crude oil and refined petroleum products regionally. The United States imports roughly 40 percent of Nigeria's crude oil and intends to import more in the future.

Currently, estimated recoverable oil is about 35 billion bbl, although the government has proposed increasing it to 40 billion. The country produces about 2.5 million bpd. The oil fields, totaling around 250, are largely located in the Niger Delta, with some bitumen reservoirs on the western side of the Niger River. These fields are located within water depths of less than 1,000 feet.

More than 95 percent of oil production is in the hands of joint ventures between major oil companies and the government-owned company the NNPC. Nigeria formed a state-owned oil company, the Nigerian National Oil Corporation, in 1970 and joined OPEC in 1971. This company, lasted until 1977, when it became the NNPC. Aside from the NNPC, Royal Dutch/Shell has been the most active oil company operating in Nigeria.

Royal Dutch/Shell is the largest producer in Nigeria, it accounts for almost half the country's daily production. The company, Shell Petroleum Development Company of Nigeria (SPDC) controls more than 100 oil fields and almost 4,000 miles of pipeline. SPDC operates two coastal export terminals, Forcado and Bonny Island, and two subsidiary companies, Shell Nigeria Exploration and Production Company Limited and Shell Nigeria Gas Limited. Nigeria represents one of Royal Dutch/Shell's largest oil reserves. ExxonMobil, the second largest company operating in Nigeria, operates in shallow-water offshore oil fields. The company also operates the Qua Iboe export terminal. ChevronTexaco operates twenty-five fields, which are mainly offshore fields, and recently began deepwater production. Its crude oil terminal is located at Escravos. Nigerian Agip produces oil from 146 wells and operates an export terminal at Brass River. Elf Nigeria produces oil from twelve onshore and offshore fields and does not operate an export terminal. In addition to the major companies, there are numerous private companies operating in Nigeria that have received licenses to explore and produce oil.

The major oil-producing companies on behalf of the NNPC, under the joint-venture arrangement, manage export terminals. In addition to the terminals previously mentioned, there are two others: Penington and Odudu. The pipelines in Nigeria are quite extensive and run from nearly border to border on the contiguous sides of the country. In recent years, they have become a target of vandalism. From the pipeline, the crude oil goes to the refineries for processing into various petroleum products. The NNPC currently operates four refineries in Port Harcourt, Warri, and Kaduna. Due to poor management and upkeep as well as sabotage, they do not run at full capacity. NNPC designed the Kaduna refinery to meet Nigeria's local demands by refining imported heavy crude oil with a high sulfur content to make petroleum derivatives such as asphalt, gasoline, and kerosene. Nigeria imports the heavy crude oil from Kuwait, Saudi Arabia, or Venezuela. In March 2003, repairs began on all the refineries. Since 2000, President Obasanjo has been pushing his plans for privatization of the downstream sector into action. In 2005 plans for the building of independent small refineries has been in the works. In addition to the four oil refineries, Nigeria's petrochemical plants, pipelines, and marketing company are due to be sold. Nigeria has also worked on expanding its natural gas industry.

While Nigeria has been slower than other oil-rich countries to develop a major natural gas industry, it is catching up. Nigeria has called for an end to flaring off natural gas by 2008. Its current natural gas reserves stand at 176 Tcf. It is estimated, however, that Nigeria flares off about 200 billion cubic feet of the 500 billion produced per day. Nigeria's natural gas is derived from oil fields; therefore, most of Nigeria's efforts have been involved in trapping and processing it. In 1999, Nigeria completed its first LNG facility on Bonny Island. Other facilities are in the planning stage.

Since the discovery of oil, Nigeria has struggled to balance nationalist interests with pragmatic economic policy. While the results are yet to be seen, Nigeria may be on the right track under its new democratically elected president, who has expressed a deep commitment to stamping out corruption, reigning in Nigeria's economy through privatization, and reducing religious-based violence.

FURTHER READING

For an overview of Nigeria's history and culture, see Toyin Falola's *History of Nigeria* (Westport, CT: Greenwood Press, 1999) and *Culture and Customs of Nigeria* (Westport, CT: Greenwood Press, 2001). See also Eghosu Osaghae's *Crippled Giant: Nigeria Since Independence* (Bloomington: Indiana University Press, 1998). For an overview of Nigeria's political economy in relationship to its oil industry, see Sarah Ahmad Khan's *Nigeria: The Political Economy of Oil* (New York: Oxford University Press, 1994) and Goddy Ikeh's *Nigerian Oil Industry: The First Three Decades (1958–1988)* (Lagos, Nigeria: Starledger Communications, 1991). For an in-depth discussion of the conflict in the Niger Delta, see Jedrej Georg Frynas's *Oil in Nigeria: Conflict and Litigation Between Oil Companies and Village Communities* (Hamburg, Ger.: Lit Verlag, 2001) and Cyril Obi's *Changing Forms of Identity Politics in Nigeria Under Economic Adjustment: The Case of the Oil Minorities Movement in the Niger Delta* (Uppsala, Swed.: Nordiska Afrikainstitutet, 2001). To read about the conflict in the Niger Delta through the eyes of an Ogoni activist, see Ken Saro-Wiwa's *Genocide in Nigeria: The Ogoni Tragedy* (London: Saros International, 1992).

Chapter 13

Norway

NORWAY
Official Name: Kingdom of Norway
Location: Western coast of the Scandinavian Peninsula
Capital: Oslo
Major Port City: Oslo
Independence: October 26, 1905 (from union with Sweden)
Population: 5 million
National Currency: Norwegian krone (NOK)
Official Language: Norwegian
Government: Prime Minister Kjell Magne Bondevik (since 2001)
OPEC Member: No
Proven Crude Oil Reserves: 10.45 billion bbl
Proven Natural Gas Reserves: 75 Tcf
National Oil Company: Den Norske Stats Oljeselskap A/S (Statoil)
Crude Oil Production: 3.2 million bpd
Natural Gas Production: 2.6 Tcf per year
Production Ranking: Seventh largest
Export Ranking: Third largest

Map 13.1
Norway

The Kingdom of Norway lies along the western coast of the Scandinavian Peninsula and shares a border with Finland and Sweden along its south-eastern border. To the north of Norway lies the Norwegian Sea, which stretches north to the Arctic Ocean, crossing into the Arctic Circle. Norway has also claimed a group of Arctic islands to the north called the Svalbard, which means "the cold coast," since 1920. Norway's terrain includes large mountains, fertile valleys, and fjords as well as a maritime territory of about 1.2 million square miles. Most Norwegians share a common ethnic and linguistic composition related to their Scandinavian neighbors. The official language in Norway is Norwegian. Norway is also home to a few minority groups, namely the Sami and the Finns, who speak their own languages. These groups live predominantly in the north, with about 40,000 Sami living as seminomadic reindeer herders that migrate between Russia, Finland, and Norway. Overall, Norway has enjoyed a tranquil history in comparison to other oil-producing countries.

Much of Norway's history and economy has been shaped by its location in relation to the sea and its neighbors. Norwegians endured a brief period of foreign occupation, when Sweden invaded Norway in 1814 and ruled during the nineteenth century. A strong nationalist movement forced Sweden to grant Norway independence in 1905. Because less than 5 percent of Norway is arable land, the Norwegian economy depends heavily on fishing, mining, and manufacturing. The country's large number of industrial workers has resulted in laborers taking a central role in politics. In the 1920s and 1930s, Norway struggled with financial crises and saw the rise of a labor movement. Between 1919 and 1939 it was estimated that a million days of work were lost annually due to strikes, which often lasted for several weeks at a time. The labor movement turned into a major political party, which controlled Norway's politics until 1981. Although Norway remained neutral in World Wars I and II, it did not escape a five-year occupation by Nazi Germany. From the 1940s to the 1980s, Norway was ruled primarily by the Labor Party, which advocates a modern form of socialism. Since the fall of the Labor Party in 1981, Norway's governments have alternated between conservative and labor-oriented leadership. The quest for an egalitarian society has always been a fundamental part of Norway's politics.[1] To protect this delicate balance, Norway decided against joining the European Union (EU) in 1994.

NORWAY'S OIL INDUSTRY

Discovery

In 1959, oil explorers found natural gas fields in the North Sea off the coast of the Netherlands. Based on this discovery, the Norwegian government concluded that natural gas could be found in their portion of the sea as well. Two major developments, however, needed to take place before Norway could become the major oil producer it is today. First, oil extraction techniques needed to advance into offshore drilling, particularly in deep water miles from a coastline. This technology advanced significantly in the 1950s. Second, international laws regarding territorial claims over the North Sea needed to be settled. In 1958, forty-six nations signed the Convention on the Law of the Sea, which gave each coastal state legal hold over its continental shelf for the purpose of exploiting its natural resources.[2] Once Norway received the go-ahead, it quickly invited foreign companies in to explore.

Norway recognized from the start that the key to maximizing its profits and control over its new industry included an unwavering commitment to government participation and strict regulations. The goal of this policy was to introduce the new industry in Norway's economy slowly without disrupting it. Norway sought to control its industry by only granting an exploration license or a production license, but not both at the same time as other oil producing countries had done. Norway launched its industry by granting exploration licenses to Phillips (which petitioned for exclusive rights and was denied) and a consortium of Exxon, Royal Dutch/Shell, British Petroleum, and CFP. The Norwegian part of the North Sea was divided into a grid system, with each block measuring several hundred square miles. The blocks were awarded over a span of several decades, with the first round in 1965 and the most recent round, the twentieth, in 2003. By the end of 1965, 278 blocks in the Norwegian North Sea had been granted. In 1969, Phillips Petroleum discovered an oil and gas field, which it named Ekofisk. Once production in Ekofisk got underway, Norway decided that implementing government regulations was not enough. They decided that a separate body connected to the government would reign in profits and regulate production more effectively.

The Norwegian government (the Storting) created a national oil company in 1972. The new company, Den Norske Stats Oljeselskap A/S (better known as Statoil), formed to participate in and oversee Norway's oil industry. In 1975 Statoil received nine blocks to develop in the North Sea. In 1977 the first pure gas field, Frigg field, was discovered. Twenty years later, offshore drilling in the North Sea saw a technological landmark, when, in 1996, a 1,500-foot-high concrete platform was erected. This platform can supposedly be seen with the naked eye from the surface of the moon.[3] North Sea production grew as major discoveries continued throughout the 1980s and into the 1990s. The first oil field in the Norwegian Sea began production in 1993 and an important gas field, the Huldra, began production in 2001.

The Industry Today

From the moment of discovery, Norway has benefited from its proximity to European markets and the fact that its entire oil and gas industry is offshore. Norway has also been quite attractive for foreign oil companies because of its political stability. Since 1975, Norway has been self-sufficient in oil. It produces approximately 3.2 million bbl of crude oil per day and holds over 10 billion bbl in reserves. In natural gas, Norway holds 75 Tcf in reserves and produces 2.6 Tcf per year. Norway is currently the world's third largest oil exporter, after Saudi Arabia and Russia. Norway is also the ninth largest producer of natural gas in the world. Much of its natural gas is associated, but recent discoveries in the Norwegian Sea have been mainly gas fields. Thus far, all the oil produced from the Norwegian fields has been the most valuable type—light with extremely low sulfur content. In fact, oil from the North Sea is used as a benchmark for all crude oils traded on the IPE in London. In addition to the North Sea, Norway also produces oil from the Norwegian Sea, and consideration has been given to opening the Barents Sea for production in the future. Norway holds only a portion of each of these claims. The North Sea is shared with the United Kingdom; the Barents Sea, with Russia. Nonetheless, Norway's holdings have made it one of the world's largest producers and exporters. Through the careful organization of a state-owned oil company, Norway has been able to manage its resources and wealth.

Unlike oil-producing countries of the past, Norway has developed a successful balance of state control while allowing foreign oil companies to operate key sectors of its industry. Until 1999, the Norwegian government owned all of Statoil and a large portion of a Norwegian oil company called Norsk Hydro. In 2001, the Storting approved the privatization of 21.5 percent of a section of Statoil's holdings. All the production contracts in Norwegian waters are joint ventures in which Statoil holds at least a 50 percent share. With its extensive involvement in Norway's industry, Statoil has positioned itself as the tenth largest oil company in the world.[4] Much of its wealth comes from the North Sea.

The oil and gas fields located in Norway's portion of the North Sea include Ekofisk, Friggs, Sleipner, Statfjord, Oseberg and Troll Vest, Troll C, Gullfaks, and Snorre. The North Sea region has accounted for 90 percent of Norway's oil and gas production since 1971. Altogether, the North Sea holds almost 900 million bbl in reserves. The Friggs and Heimdal fields are primarily natural gas fields. The Troll fields hold Norway's largest natural gas reserves. The Statfjord field is Norway's largest North Sea oil field, which was discovered by Mobil in 1974. Today this field is showing signs of decline. Within the past few years, oil reserves in the Norwegian Sea are catching up to those in the North Sea, with more than 700 million bbl of reserves. Production in the Norwegian Sea began in 1993. The Norwegian Sea's oil fields include Draugen, Heidrun, and Norne. These fields account for 25 percent of Norway's oil production. The North Sea and Norwegian Sea fields are connected by a network of pipelines to terminals in Norway as well as the United Kingdom and mainland Europe. Deliveries of natural gas run through Europipe I from a platform in the North Sea to Dornum, Germany, and through Europipe II from Kårstø, Norway, to Dornum, Germany. Both lines came onstream in the second half of the 1990s. More pipeline plans should develop, as two major natural gas fields in the Norwegian Sea have been discovered in the past couple of years. Plans for exploration have also been in the works for the Barents Sea.

The Barents Sea oil fields, for now, account for a very small percentage of Norway's total production. Norway has moved cautiously toward new exploration and production projects close to the Russian border because of the environmental sensitivity of the region and Russia's shared claim to the territory. Environmentalists have lobbied Norway to protect, not develop, the Barents Sea because it is a fragile environment that cannot withstand oil spills. Until 2003, Norway refused major exploration in the region. Russia, on the other hand, has less strict environmental standards and is eager to begin. Preliminary steps were taken between the two countries in 2002 with an agreement to discuss exploration and the sharing of potential oil reserves. Two years later, the countries signed a letter of intent regarding oil pollution action and security for tankers in the Barents Sea. With the chance that production in the Barents Sea may not materialize, Norway has been preparing for the possibility that its years as a major crude oil producer are ending.

While gas and oil production have increased since their discovery in the 1960s, Norway has taken precautionary measures to ensure that when the oil runs out, the country is prepared. The North Sea fields are maturing, with no new discoveries in the region likely. The exhaustion of Norway's oil reserves has been calculated to take place in the first half of the twenty-first century. In preparation, Norway has set up a Petroleum Fund, which the government pays into annually. This source of money was developed to supplement the decline in oil revenues that will take place.

Norway has also been revered for its environmental sensitivity when it comes to the impact of oil. In 1982, a marine biologist discovered cold-water corals in the Norwegian North Sea while surveying for an underwater pipeline project known as the Snøhvit project. Without much debate, Norway drafted plans to reroute the pipeline, which is expected to be completed in 2005, around the Tisler Reef. When completed, the Snøhvit project will be the largest underwater LNG endeavor in the world.

The pipeline will carry natural gas from fields in the Barents Sea to the Snøhvit plant on Melkoya Island. From the LNG plant, the gas will be shipped to Europe and North America. On the other hand, Norway has received a great deal of criticism from environmentalists on its policies toward whaling. While an international moratorium was passed in 1986, Norway continued the practice, essentially because for centuries Norway's economy has depended heavily on its fishing industry. With the country's oil industry in potential decline, the reliance on fishing may only increase.

FURTHER READING

A surprisingly small number of books are available in English on the history of Norway. Fortunately, the recently launched *Cambridge History of Scandinavia* series offers a multi-volume overview of the region's history. As of publication, however, only the first volume has been published: Knut Helle, ed., *Cambridge History of Scandinavia*, Vol. 1, *Prehistory to 1520* (Cambridge: Cambridge University Press, 2003). See also Thomas K. Derry's *A History of Modern Norway, 1814–1972* (Oxford: Clarendon Press, 1973). In comparison, a larger number of books are available on Norway's oil industry. Several works compare Norway's offshore activities in the North Sea to those of its neighbors. See Øystein Noreng's *The Oil Industry and Government Strategy in the North Sea* (London: Croom Helm, 1980) and Brent F. Nelsen's *The State Offshore: Petroleum, Politics, and State Intervention on the British and Norwegian Continental Shelves* (New York: Praeger, 1991). For a discussion comparing Norway's and Mexico's handling of their oil industries during the 1970s and 1980s, see Ragaei El Mallakh's *Petroleum and Economic Development: The Cases of Mexico and Norway* (Lexington, MA: Lexington Books, 1984).

Chapter 14

Russia

RUSSIA
Official Name: Russian Federation
Location: Northern Asia
Capital: Moscow
Major Port City: Novorossiysk
Independence: August 24, 1991 (from the Soviet Union)
Population: 145 million
National Currency: Ruble (R)
Official Language: Russian
Government: President Vladimir Vladimirovich Putin (since 1999)
OPEC Member: No
Proven Crude Oil Reserves: 60 billion bbl
Proven Natural Gas Reserves: 1,680 Tcf
National Oil Company: Rosneft (oil) and Gazprom (natural gas)
Crude Oil Production: 9 million bpd
Natural Gas Production: 22 Tcf per year
Production Ranking: Third largest
Export Ranking: Second largest

Map 14.1
Russia

Map 14.1 Russia

RUSSIA
····· River
● Oil field

miles
0   500

Arctic Ocean

Wrangel Island

Franz Josef Land

Svalbard (Norway)

New Siberian Islands

Severnaya Zemlya

Laptev Sea

Eastern Siberia

Sea of Okhotsk

Kuril Islands

Sakhalin Island

Novaya Zemlya

Kara Sea

Barents Sea

NORWAY

Arctic Circle

Oslo

SWEDEN

Stockholm

FINLAND

Murmansk

Helsinki

Tallinn

ESTONIA

Saint Petersburg

LITHUANIA

Vilnius   Riga

BELARUS

Mnsk

Kiev   UKRAINE

MOLDOVA

Chisinau

Moscow

Kazan

Volga

Samara

Rostov

Novorossisk

GEORGIA

Tbilisi

ARMENIA

Yerevan

AZERBAIJAN

Baku

Caspian Sea

Ashgabat

TURKMENISTAN

Ural Mountains

R U S S I A

Western Siberia

Ob

Omsk

Aqmola

KAZAKHSTAN

UZBEKISTAN

Tashkent

Bishkek

KYRGYZSTAN

Dushanbe

TAJIKISTAN

Central Siberia

Lake Baikal

Angarsk

Ulanbataar

MONGOLIA

CHINA

Daqing

N. KOREA

Vladivostok

Nakhodka

Sea of Japan

JAPAN

Pyongyang

Seoul

S. KOREA

Russia is the world's largest country, but it is not its sheer size that makes it remarkable. After years of being the center of a communist empire as well as one of the world's largest producers of oil, Russia has developed a unique place for itself in the world. Since the breakup of Soviet Union in 1991, Russia has worked toward opening its doors to new political alliances and foreign investment.

Russia's territory extends halfway around the globe. Its Ural Mountains traditionally divide Europe from Asia. To the west of the Urals, Russia borders the Ukraine, Finland, Norway, Estonia, Latvia, and Belarus. The landscape and climate in this region resemble those of continental Europe, with the Volga and Don rivers flowing north to south and the Caspian and Black seas located to the south. Between the two seas lie the Caucasus Mountains, which create a natural border with Georgia and Azerbaijan. The majority of Russia's population lives in this region and works as farmers. To the east of the Ural Mountains, Russia shares a border along the south with North Korea, Mongolia, China, Kazakhstan, Georgia, and Azerbaijan. Here the terrain resembles that of Russia's neighbors, with uplands and mountains, although most of Russia's terrain is relatively flat. Along the Mongolian border lie the Tanna-Ola Mountains as well as the Sayan and Yabonovyy mountains. In the northern region, Russia's climate and terrain become subarctic tundra, which extends north to the Arctic Ocean. This area is known as Siberia and occupies much of Russia's landscape. On the eastern frontier lie the Cherskiy Mountains, which merge with the Stanover Mountains. East beyond the mountains is the Sea of Okhotsk, which connects to the North Pacific Ocean. Because Russia covers the whole of north Asia and part of Europe, its population is a diverse mixture of people.

The Russian Federation is home to more than sixty different nationalities, which speak a variety of languages, aside from Russian, its official language. Over the years, political and ethnic unrest have plagued Russia's population. Since the 1800s Muslim separatists in Chechnya, for example, have demanded independence from Russia; they received partial autonomy within the federation in 1991 with that fall of the Soviet Union. Chechnya is a small republic of mainly Muslim people near the Caucasus Mountains. In 1994, tension between Russia and the Chechens escalated when Russia sent in troops to crush an independence movement calling for full independence. Although the situation has improved in recent years, there is no guarantee of continuing tranquillity. In an attempt to maintain a level of autonomy, representation, and infrastructure today, Moscow runs Russia as a federation.

Within the federation there are several republics representing various groups. The country, overall, is comprised of eighty-seven administrative divisions that include autonomous republics (such as Chechnya and the Jewish province of Yevrey), and federal cities (Moscow and Saint Petersburg). These regions freely elect their governments while participating in the national government. The reason for such a complicated political system stems from a long history of Russian territorial and political expansion and contraction, which began in the early twentieth century.

Russia has an interesting but turbulent history full of famous leaders and dramatic economic and social change. In the 1910s, the rising discontent of Russia's working class generated radical political thought. The country's social and economical ravishment caused by World War I, resulted in a revolution, which occurred in October 1917. The revolution began with the Bolsheviks, led by Vladimir Ilyich Ulyanov (code-named Lenin), staging a coup d'état, which marked the beginning of more than seventy-five years of communist rule. The transition into a new political structure, however, was not without its problems.

A civil war broke out from 1917 (after the revolution) until 1922. The war involved a series of military clashes between the Bolsheviks, referred to as the Reds, who took power and its opposition, commonly referred to as the Whites, who hoped to restore Russia to its pre-revolution tsarist rule. The Whites received support from Britain, France, the United States, and Japan. Despite this support, the Bolsheviks succeeded in establishing power throughout Russia and its neighbors, forming the Soviet Union. Under Lenin's rule (1917–1924), the government promised to end Russia's involvement in the war and to implement land reform. The Union of Soviet Socialist Republics (USSR) formed the world's most geographically expansive sovereign state. It functioned as a federation of fifteen republics. After Lenin's death, Joseph Stalin (ruled 1924–1953) took over and implemented plans to solidify the Soviet Union's control. In 1928 he introduced a five-year plan, which included major industrialization and the collectivization of Russia's peasants. Under Stalin, millions died from starvation and systematic government brutality, which included the forced placement of people in Gulags (Russian concentration camps). After World War II, Russia became involved in the cold war standoff with the United States. Both sides were poised for mutual destruction. For example, In 1962, under Nikita Khrushchev (ruled 1958–1964), the Soviet Union averted a hot war when the United States contested the presence of Soviet nuclear missiles in Cuba. Acts of repression toward its neighbors (an occupation in Afghanistan from 1979 to 1989) and its own people, particularly under Leonid Brezhnev (ruled 1964–1982), did little to improve the image of the Soviet Union in the eyes of the international community, and especially the United States.

In the late 1980s, the Soviet economy began to collapse and rebellions within the union increased. The Soviet Union under Mikhail Gorbachev, who served as general secretary from 1985 to 1991, went through dramatic changes. Communist power declined and independence movements picked up strength. An attempted coup erupted in August 1991, which led to the collapse of the Communist Party and subsequently the fall of the Soviet Union in December of that year. Between March 1990 and December 1991 several regions became independent (Armenia, Azerbaijan, Belarus, Estonia, Georgia, Kazakhstan, Kyrgystan, Latvia, Lithuania, Moldavia, Tajikistan, Turkmenistan, Ukraine, and Uzbekistan). With the exception of Estonia, Latvia, Lithuania, and Turkmenistan, all the former Soviet republics joined the Commonwealth of Independent States in December 1991. The Soviet Union officially dissolved what remained of the former Soviet Union, becoming the Russian Federation. As president of the new Russian Federation, Boris Yeltsin (president 1992–1999) took the initiative of formally ending the cold war, which had lasted more than forty years. At the end of 1999 he resigned, and Vladimir Putin, who was Yeltsin's prime minister, took his place.

Since the dissolution of the Soviet Union, the Russian Federation continues to develop economically and to open its doors to international companies and trade agreements. Progress, however, has been slowed by high inflation and Russian mafia control.[1] Under Putin, the Russian Federation has attempted to improve the lives of its people and expand its industrial base. Putin's recent firing of his government, however, has raised concern over the country's political future. Nonetheless, Putin has placed particular emphasis on Russia's oil and gas industry. The sheer size of the country as well as its severe climate create challenges for the Russian Federation, but Russians remain optimistic.

RUSSIA'S OIL INDUSTRY

Discovery

The discovery of Russia's oil begins with travelers witnessing oil seeps near the Caucasus Mountains in southwestern Russia in the early nineteenth century. Production in the area remained small until a family of engineers found the Baku fields. Today these fields fall within Azerbaijan's borders on the coast of the Caspian Sea, but prior to Azerbaijan's independence in 1991 they fell under the control of the Soviet Union. In the 1860s, two brothers, Robert and Ludwig Nobel, focused on oil production in the Baku region. At the same time, the Rothschilds from Germany were also involved in expanding their kerosene trade. In 1891 Marcus Samuel, the founder of the Shell Transport and Trading Company (later to become Royal Dutch/Shell) signed a contract with the Rothschilds to buy their Russian oil. In 1911 Royal Dutch/Shell purchased the Rothschilds' holdings for 2.75 million Russian rubles.[2] In 1916 the Grozny fields were developed, increasing Russia's output to 15 percent of the world's production. By 1916 the Nobels produced more than 30 percent of Russia's crude oil and supplied 60 percent of it for domestic consumption.[3] They purchased a concession, received financial assistance from the Rothschild family, and borrowed technology from the United States to drill deep wells. Their discoveries marked the beginning of commercial production. By the turn of the century, Russia contributed more than 30 percent to the world's oil production, with the Nobel brothers controlling about one-third of it. Until World War I, Russia held a leading position in the development of oil-processing technology.[4] Royal Dutch/Shell ferried the oil to Western Europe, as Russia's oil industry thrived until the Russian Revolution.

The rise of the Communist Party brought dramatic changes to Russia's oil industry. The October Revolution of 1917 inspired workers' uprisings in the oil fields against low wages and harsh working conditions. In 1919 Azerbaijan took advantage of the political unrest to declare sovereignty over the Baku fields. That same year SONJ made an agreement with the Azerbaijan government to purchase undeveloped land for exploration in the Baku region.[5] Amidst the chaos, foreign oil companies rushed into Russia hoping to collect concessions at reduced rates. The Nobel brothers sold much of their operations to SONJ (today ExxonMobil) to build an alliance in 1920. That same year, however, the Soviets seized foreign holdings and played the major companies off each other while building a national oil industry. In retaliation, the major oil companies severed connections with Russia. Soviet planners began to accelerate its oil industry by exploring new areas. In the 1930s, they discovered oil in a region near the Volga River west of the Ural Mountains. The first oil refinery, called Omsk, which is one of the worlds largest, went onstream in 1955. By 1959, Russia had replaced Venezuela as the world's largest oil producer at 300,000 bpd.[6] In the 1960s, the first discoveries in western Siberia were made. In 1965, Russia's largest field, Samotlor, was discovered. Since then numerous fields, including the recent discoveries on Sakhalin Island in the north Pacific Ocean, have placed Russia among the world's top oil producers.

The Industry Today

Since the fall of the Soviet Union in 1991, Russia's oil industry has already gone through a number of changes. Privatization, increase in production, new exploration plans, and company mergers have all taken place within a few short years. Russia is currently the largest non-OPEC producer of oil in the world, and Russia also holds the world's largest natural gas reserves at 1,680 Tcf and produces 22 Tcf per year. In September 2003 Russia's crude oil output exceeded Saudi Arabia's.[7] Russia's oil production totals 8 million bpd. At the same time, the development of Russia's oil industry is difficult because most of Russia consists of subarctic conditions and expanses of virtually untouched terrain. For this reason, transporting equipment and setting up housing for oil workers and communication networks are costly. Between the 1970s and the 1990s, Russia's planning for its oil industry was short-term. Poor extraction, processing, and transporting practices have caused serious dilapidation and environmental damage over the years. At the time, the Russian oil companies received little incentive to concern themselves with the long-term preservation of wells. Harsh climate, remote locations, and lack of capital have made the exploration of new wells in Russia, particularly in Siberia, difficult and slow. Today Russia's oil companies and government have a new set of goals for the industry that include foreign investment, conservation, production increases, and greater global trade links.

The transformation of Russia's oil industry included the breaking down of national companies into privatized units. Presidential Decree No. 971 of 1995 called for the transfer of state-owned enterprises and associations into joint-stock companies. This formed a number of giant oil companies to control the country's vast natural and industrial resources. It also invited foreign companies such as British Petroleum, ExxonMobil, and Royal Dutch/Shell. These companies, through joint ventures, have contributed to the modernization, revitalization, and exploration of new fields in Russia.

Today Russia's oil industry is dominated by a handful of large joint-stock companies—Lukoil, Yukos, TNK-BP, and Rosneft. Lukoil is Russia's largest oil company, with several subsidiary companies that not only produce oil, but also market it and manage retail outlets. It controls 19 percent of Russia's oil production and operates refineries in Romania, Bulgaria, and the Ukraine. In 2000, Lukoil was said to have reserves that matched or exceeded those of ExxonMobil.[8]

Yukos is Russia's second largest oil company. Its primary activities include oil and gas exploration, production, refining, and marketing in western Siberia. Recently, the CEO of Yukos, Mikhail B. Khodorkovsky, and his partners have been in the public eye for underhanded dealings with banks and Yukos's shareholders. They are accused of tax evasion. As Russia's richest man, Khordorkovsky holds a great deal of political sway. It has been said that Khodorkovsky gave financial support to political opposition parties.[9] For this reason, Khodorkovsky's imprisonment has raised suspicion. Many view the Russian government's accusations as a thinly disguised effort to punish Khodorkovsky by forcing him to sell of parts of Yukos and spend up to ten years in prison.

Russia's third largest producer, TNK-BP, formed out of a September 2003 merger between OAO TNK, British Petroleum, and SIDANCO, with BP managing the company. TNK-BP operates fields in the Tyumen region of western Siberia, the Irkutsk region of eastern Siberia, and the Ural Mountains.

In addition to Russia's independent joint-stock companies, the Russian government has its own state-owned company, Rosneft, which produces oil from fields in western Siberia and the island of Sakhalin. Collectively, these companies make up the large newly independent oil companies based in Russia.

Russia holds some of the largest oil fields in the world, which total 60 billion bbl of proven reserves. Russia produces oil in six major regions. The north and northwestern coasts of the Caspian Sea, known as the North Caucasus and Astrakhan area, include more than 100 fields. The Timan-Pechora oil fields total more than 160 and are located directly to the west of the Ural Mountains and just south of the Arctic Ocean. West of the Ural Mountains are also the Volga-Ural and Saratov-Volgograd areas, which contain more than 600 oil fields. To the east of the Ural Mountains in western Siberia is the Tyumen/Ob, which includes more than 360 oil fields. In the 1980s, this area represented the center of Russia's oil industry. The final region includes Sakhalin Island, which contains more than twenty-five fields. Oil production on Sakhalin Island is relatively new and is divided into two development groups. The first is led by ExxonMobil in a consortium with several smaller companies. The consortium began drilling in May 2003 and expects production to begin in 2005. The oil produced will be shipped to the Russian port of De-Kastri.

Russia's oil fields have produced oil bound not only for export, but also for internal refining and domestic sales. The second group, led by Royal Dutch/Shell in a consortium with Mitsubishi and Mitsui, focuses more on natural gas.[10]

In each of the six producing regions, Russia has several refineries and access to a large web of pipelines. Overall, Russia has forty-two oil refineries, many of which process crude oil with outdated methods and are in need of major repairs. The oldest and largest oil refinery, Omsk, is located near the Kazakhstan border west of the Ural Mountains. Most of Russia's refineries are located west of the Urals near the Black and Caspian seas. The refineries located east of the Ural Mountains primarily follow pipelines that run along the southern border of the country. One of particular importance is the Angarsk, which is located just north of Mongolia along a pipeline. Overall, the pipeline network in Russia is fairly extensive, with plans for expansion under way.

Russia's pipelines connect remote fields to central lines that carry oil and gas to refineries or terminals for export. Russia has three major export lines—Northern and Southern Druzhba and Adria. The Druzhba lines connect the Volga-Ural region to the Ukraine in the south and Germany in the north. The Adria was completed in 1979 but has been under repair since the 1990s. In addition to fixing existing pipelines, the Yukos has also built new pipelines. The Baltic pipeline, completed in 2001, connects fields in western Siberia south to the Baltic Sea. A proposal has been made for a Murmansk pipeline, which would connect the Timan-Pichora fields to the Murmansk port on the Barents Sea. Plans are also in the works for a pipeline extending off Southern Druzhba to the Adriatic Sea.

The largest plan so far, however, has been a pipeline plan to connect Russia to Asian markets. While an agreement was signed for the Angarsk-Daqing pipeline by China National Petroleum Company and Yukos in 2003, construction has not moved forward. At the time of publication, Russia remained interested in the pipeline to China, but also became interested in building the Angarsk-Nakhodka pipeline, which would run oil from eastern Siberia's fields to the Sea of Japan. Nonetheless, Russia has made agreements to increase its exports by railroad and tanker to Asia.

Most of Russia's exports leave from terminals west of the Ural Mountains. On the Black Sea are two terminals—the Novorossiysk and the Tuapse. On the Baltic Sea is the Primorsk and on the Sea of Khotsk are the Okha and the Korsakov terminals. Russia also uses ports in Lithuania, Ukraine, and Latvia. In addition to Russia's use of extensive pipelines and terminals for oil, it also transports and trades natural gas.

Russia produces the most natural gas in the world. With more than 1,680 Tcf in proven reserves, Russia has plenty of gas to export after setting aside about half of it to meet domestic needs. Russia's natural gas comes from associated gas collected from oil fields and from natural gas fields, which are primarily located in western Siberia. In eastern Siberia and along the Sea of Okhotsk, natural gas fields have also been discovered.

Russia's natural gas industry is operated by the largest natural gas company in the world, Gazprom. The joint-stock company is responsible for Russia's natural gas domestic sales and exports. The Russian government holds a 38 percent share in the company. Almost all of Gazprom's natural gas exports go to Europe, which constitutes 25 percent of Europe's imports. Gazprom operates six gas pipelines (Bratrstvo, Soyuz, Northern Lights, Volga/Urals-Vyborg, Yamal, and Blue Stream), most of which pass through the Ukraine to Europe. Gazprom has several plans to expand the pipeline network. Natural gas is expected to be piped southward from Sakhalin Island to Japan via a proposed pipeline beginning in 2008. Plans for an LNG project by Royal Dutch/Shell, Mitsubishi, and Mitsui at Sakhalin Island have also been proposed. Trade contracts with Japan have been made and exports are expected to begin in 2007. Gazprom has also signed a contract with South Korea's state-owned Korea Gas Company and the Chinese National Petroleum Company to run a pipeline from Russia's Kovyota gas field to China and South Korea. Exports are expected to begin in 2008. While much of Gazprom's expansionist activities are international, it also operates on a national level.[11]

On the domestic front, Gazprom finds itself in a political bind. As a legal monopoly it is subject to the government's demands. A gas law in 1999 declared that Gazprom must supply Russia's natural gas market at regulated prices. Thus, the company is required to sell natural gas at prices lower than the market price and production costs. The Russian government, however, recognizes this problem, and promised to free natural gas prices around 2010. Overall, the Russian government recognizes that although many changes have been made to its economy, many more are to come. The country, however, does not doubt its potential to become the world's largest oil- and gas-producing country.

FURTHER READING

For an excellent overview of Russia's history, see Paul Duke's *A History of Russia: c. 882–1996* (Durham, NC: Duke University Press, 1998) and Robert Service's

A History of Modern Russia: From Nicholas II to Putin (London: Penguin Books, 2003). See also Charles E. Ziegler's *History of Russia* (Westport, CT: Greenwood Press, 1999). For a survey of not only Russia's history, but also politics and society, see Nicholas V. Riasanovsky's *A History of Russia*, 6th ed. (New York: Oxford University Press, 2000). For a brief overview of Russia's history and culture, see Sydney Schultze's *Culture and Customs of Russia* (Westport, CT: Greenwood Press, 2000). Although no one-volume book covers Russia's oil and gas industry, David Lane's *Political Economy of Russian Oil* (Lanham, MD: Rowman and Littlefield, 1999) provides a collection of essays. For information on Russia's economy since the 1990s, which addresses the formation of Russia's private oil companies, see Marshall I. Goldman's *The Piratization of Russia* (New York: Routledge, 2003).

Chapter 15

Saudi Arabia

SAUDI ARABIA
Official Name: Kingdom of Saudi Arabia
Location: Arabian Peninsula
Capital: Riyadh
Major Port Cities: Jedda, Dammam, Yanbu
Independence: September 23, 1932 (unification of the kingdom)
Population: 26 million
National Currency: Saudi riyal (SAR)
Official Language: Arabic
Government: King Fahd ibn Abd al-Aziz al-Sa'ud (since 1982)
OPEC Member: Since 1960
Proven Crude Oil Reserves: 262 billion bbl
Proven Natural Gas Reserves: 235 Tcf
National Oil Company: Saudi Aramco
Crude Oil Production: 10.5 million bpd
Natural Gas Production: 2 Tcf per year
Production Ranking: Largest
Export Ranking: Largest

Map 15.1
Saudi Arabia

Saudi Arabia holds the unique distinction of not only being the largest oil producer in the world, but also being geographically the largest country in the Middle East. Occupying almost the entire Arabian Peninsula, Saudi Arabia is considered a giant among its neighbors. It also has the distinction of being the custodian of the two most holy Islamic cities of Mecca and Medina. Muslims from all over the world travel to these holy places. Mecca is the birthplace of the Prophet Muhammad and the Islamic faith and Medina is the Prophet's burial place. Saudi Arabia shares a border to the north with Jordan, Iraq, and Kuwait, and to the south it shares a border with Yemen, Oman, and the United Arab Emirates. Saudi Arabia's coastline touches two major waterways—the Red Sea and the Persian Gulf.

Saudi Arabia is highly dependent on its petroleum industry, and a look at the country's landscape explains why. The country lacks permanent rivers and is bordered on the east and west by large bodies of saltwater. Saudi Arabia holds only two percent arable land, which is threatened by desertification. The remainder of the country includes rolling hills of sand with patches of grassland along the coasts. In the north lies the Nafud desert where distinct reddish sand forms into large dunes. The largest desert in the world, the Rub' al-Khali (or Empty Quarter), is located in the south. This expansive desert receives rain less than every ten years. Saudi Arabia also has small mountain ranges, such as the Asir Tihamah, which follow the Red Sea. Despite the rough terrain, Saudi Arabia boasts a long history of civilizations migrating and settling across the Arabian Peninsula.

The history of Saudi Arabia is largely connected to its geography. Life in the desert fostered independence and adaptability to the harsh terrain. Because of the large size of the Arabian Peninsula in relation to its sparse population, clans in Saudi Arabia lived in relative isolation. Some groups settled near an oasis and took to farming, while others became nomadic. Only along the coast did people engage in trade and commercial travel. These lifestyles protected the people for centuries from outside invaders.

The arrival of Islam and the expansion of the Al-Saud family, however, dramatically changed the history of Saudi Arabia. Islam, which began with a divinely inspired merchant, Muhammad, in 622 B.C.E., spread through Saudi Arabia quickly. Muhammad traveled north to Medina from Mecca, bringing the religion with him. When he returned to Mecca in 630 B.C.E., he brought an army of converts with him and seized the city. Mecca remained the center of Islamic power until it was moved to Baghdad in 661 B.C.E. At the time of Muhammad's death, in 632 B.C.E., Islam had become the predominant religion of Saudi Arabia. At the same time that the principles of Islam spread and acted as a unifier, different interpretations of the Koran developed that created rifts among Muslims. Divergent doctrines became popular throughout the Middle East. One prominent group, the Wahhabi, contributed to the formation of Saudi Arabia.

A Muslim scholar, Muhammad ibn-Abdul-Wahhab, founded the Islamic fundamentalist Wahhabi movement in the eighteenth century. The movement called for a return to traditional Islam. Facing persecution, he found protection in a town ruled by the Al-Saud royal family, which ruled much of the Arabian Peninsula. At the time the Arabian Peninsula was made up of four major regions—Asir, Hasa, Hijaz, and Nejd. The Al-Saud family's control placed them in direct conflict with the Ottoman Empire as well as the British Empire, both striving to expand over the region. By the late nineteenth century, however, the Ottomans triumphed and the Al-Saud family went into exile near Kuwait.

At the close of World War I, the Ottoman Empire collapsed, leaving Saudi Arabia split into two dynasties: the Al-Saud's in the east and the Hashimites in the west. The Al-Saud family ultimately defeated the Hashimites through an alliance with the British, who claimed much of the Middle East through the carving up of the Ottoman Empire. After years of negotiation with the British, the Al-Saud family prevailed as sovereign and independent rulers. Throughout the 1920s and 1930s, the royal family continued to gain political strength through marriages with important families in the country. In 1932, Saudi Arabia declared itself an Islamic state, using the Koran as its constitution. This commitment to Islam continues today. Saudi Arabia's commitment to Islam, however, has not isolated Saudi Arabia from its neighbors or the West.

Today, Saudi Arabia grapples with a unique situation. Saudi Arabia is the world's largest producer and exporter of petroleum, bringing to the country large amounts of oil revenue. At the same time, the country faces economic difficulties of high unemployment coupled with high government spending. Aside from oil, Saudi Arabia's economy depends on pilgrimage traffic, which brings in millions of people a year. This dependence on oil has created a long-term challenge for Saudi Arabia because the country has little other industry or agriculture to speak of. Saudi Arabia has pushed forward development projects, however, and attempted to attract foreign investment. This is not to say that Saudi Arabia's population suffers, because, overall, Saudi Arabia's population enjoys a high standard of living, with most modern amenities.

SAUDI ARABIA'S OIL INDUSTRY

Discovery

Exploration for commercial amounts of oil began in the nineteenth century. Until the 1920s, oil companies remained skeptical about oil existing in Saudi Arabia and instead focused on Iraq and Iran. In 1933, the king of Saudi Arabia granted a concession to an engineer from New Zealand, Major Frank Holmes. He worked closely with SOCAL in negotiations with Saudi Arabia until SOCAL expanded and brought in partners.

The 1934 concession gave SOCAL the exclusive right to every aspect of oil production in the country for sixty-six years. Saudi Arabia, in exchange, received gold as payment. Shortly after SOCAL received the concession, CASOC formed and took over the concession. In 1936 Texaco joined the company. Five years after signing the concession agreement CASOC found commercial quantities of oil in Dhahran, near the Persian Gulf. The field was a textbook example of a geological dome that contained oil. In 1939, the first shipment of oil left Saudi Arabia. In 1944 these companies, including CASOC, formed the Arabian American Oil Company (Aramco), which controlled Saudi Arabia's oil industry until the 1970s. Saudi Arabia's largest oil field, Ghawar, was discovered in 1948 and accounts for more than half the country's production.

In addition to Saudi Arabia's oil industry within its borders, it also claims oil in the Saudi-Kuwaiti Neutral Zone, which was established in the 1930s by the British due to political conflict. In 1953 the first oil field in the neutral zone, Wafra, was discovered.[1] In 1969 the area was divided, with both countries receiving an equal share of the oil industry. The neutral zone has 5 million barrels of proven reserves, which, given its size, is remarkable. Since its development, the Saudi-Kuwaiti Neutral Zone has flourished, with a major oil field that produces almost 600,000 bbl of oil per day. Saudi Arabia set up a similar arrangement with Bahrain as well. Both countries jointly operate the Abu Safa oil field, which produces about 140,000 bpd. Through its own production as well as the joint ventures, Saudi Arabia has accumulated an enormous oil industry.

Industry Today

With 262 billion bbl of proven oil reserves, Saudi Arabia is the largest producer and exporter in the world and will most likely hold this title in the future. The country produces 10.5 million bpd, which is roughly one-tenth of the world's needs. Much of the demand for Saudi Arabia's oil stems from the country's ability to offer a variety of crude oils, from heavy to light. The light crude, which makes up 65 percent of Saudi's oil, comes from onshore fields. The medium and heavy crudes come predominantly from offshore operations. On the world oil market, Saudi Arabia offers several major crude oils, such as Arabian Light, Arab Extra Light, Arab Super Light, Arab Medium, and Arab Heavy. In producing these fields, Saudi Arabia works closely with foreign oil companies such as Chevron-Texaco, BP, and ExxonMobil. While maintaining a close alliance with Western nations, Saudi Arabia also holds a dominant position in OPEC as the swing producer country, which cuts and increases production in its state-owned company as needed to protect the price of oil on the world market. With so much oil, the country feels little direct impact from any slight change to its rate of oil production, unlike smaller and more unstable member countries in OPEC. A country with so much oil, must have good management of its industry.

Saudi Arabia's oil industry is owned and operated by complex arrangements that balance foreign investment and state control. All of Saudi Arabia's oil production is overseen by the Supreme Petroleum Council under supervision of the king. Because the country so heavily depends on its oil industry, the king is the ultimate decision maker regarding Saudi Arabia's oil. Presently, however, the crown prince has been standing in for him because the king is extremely ill. Under the Supreme Petroleum Council's authority, several companies operate in the country as well as in the neutral zone and Bahrain. The three major companies include Saudi Aramco, Saudi Texaco, and Japan's Arabian Oil Company (AOC). Saudi Aramco serves as the state oil company and produces 98 percent of the kingdom's crude oil and natural gas both onshore and offshore. Since it is the king's oil company, it has been described as the world's largest family business. The Saudi Aramco company formed out of the foreign-owned Aramco. In 1976, Saudi Arabia decided to nationalize Aramco, but instead of outright expropriation, the king arranged a gradual handover, which was completed in 1980. A name change to Saudi Aramco in 1988 reflected the new ownership. The original Aramco still operates in Saudi Arabia, but only as a provider of equipment and technological assistance. Through this same royal decree, the kingdom also took a share in Texaco's operations, forming Saudi Texaco.

Saudi Arabia's additional production interests outside of the country also operate under the Supreme Petroleum Council. The Saudi-Kuwaiti Neutral Zone is operated by the AOC and Saudi Texaco. In Bahrain, Saudi Aramco acts as the main oil producer. In addition to profit sharing, Saudi Arabia also donates some of its revenues back to Bahrain. While these operations are important to Saudi Arabia, the kingdom depends on the enormous fields within the country.

The majority of Saudi Arabia's oil fields are located west of Riyadh, particularly along the Persian Gulf, with a few on the Saudi side of the Persian Gulf. Within Saudi Arabia proper, there are eighty oil and gas fields. More than half the country's total reserves are held in only eight fields. Saudi Arabia, and the world's, largest onshore oil field is Ghawar. This one field holds 70 billion bbl of proven reserves, which exceeds many oil-producing countries' total proven reserves. As the field matures, however, it is becoming costly to maintain. Safaniya is the world's largest offshore oilfield, with reserves estimated at 19 billion bbl.

Although Saudi Arabia produces the most oil in the world, it has a surprisingly small number of pipelines in operation. Currently, Saudi Arabia operates one major oil pipeline that transports oil from fields to refineries and export terminals. The East-West Crude Oil Pipeline, also known as Petroline, was built by Mobil in 1981 and expanded in 1987. Operated by Saudi Aramco since the 1980s, the pipeline transports light oils along an east-west route toward refineries and export terminals on the Red Sea. Saudi Arabia had two other pipelines running, but political instability forced them to close. The Trans-Arabian Pipeline (Tapline), which opened in 1950, travels from fields along the Persian Gulf northeast toward Jordan, but has been closed since 1976. Saudi Arabia also shares a pipeline with Iraq that follows Petroline and then breaks away, traveling north and crossing a mutual border near the Red Sea. That pipeline, however, has been closed since the Gulf War in 1990. Since then Saudi Arabia has claimed ownership of the pipeline, even though no oil flows through it. A new oil pipeline that runs from the Empty Quarter to Yemen has been discussed, but no formal arrangements have been made.

The majority of Saudi Arabia's crude oil travels through the Petroline to the Abqaiq crude-oil-processing plant. Saudi Arabia also has several export terminals and, in most cases, refineries on the same site. On the coast of the Red Sea are located Jedda and Rabigh, which have refineries operated by Aramco. The Yanbu export terminal on the Red Sea has two refineries—one operated by Saudi Aramco and the other by Saudi Aramco and ExxonMobil. On the coast of the Persian Gulf is the Ras al-Ju'aymah export terminal. Saudi Arabia has two offshore facilities in the Persian Gulf, which include Zuluf and Ras Tanura, which is the largest in the world. The Ras Tanura also has a refinery onsite operated by Saudi Aramco. In the capital city, Riyadh, Saudi Aramco also operates a refinery. In the Saudi-Kuwaiti Neutral Zone, the AOC operates the Ras al-Khafji refinery, which also serves as the region's export terminal. After the oil leaves Saudi Arabia's export terminals, it is not necessarily out of Saudi Arabia's control, because the country owns or shares oil storage facilities and refineries in Asia, Europe, and North America. In the United States, Saudi Arabia is a partner in three refineries. In Asia, Saudi Arabia owns one in the Republic of Korea and one in the Philippines.

For the future of its oil industry, Saudi Arabia has looked for ways to expand its petrochemical and natural gas industries. The partially state-owned Saudi Arabian Basic Industries Corporation expanded the coun-

try's petrochemical industry by connecting new facilities to existing refineries such as Yanbu. Another petrochemical plant went online in 2004
in Jubail, making it the world's largest. Saudi Arabia has also begun to expand its natural gas industry. Replacing oil consumption with natural gas
has been a high priority for Saudi Arabia. Saudi Arabia holds 235 Tcf of
natural gas. It produces 2 Tcf per year. The majority of Saudi Arabia's natural gas (associated and non-associated) comes from the Ghawar field. In
2003, Saudi Arabia began production from a new gas well in the Abqaiq
field in the Ash Sharqiyah (Eastern Province). In the Saudi-Kuwaiti Neutral Zone, the AOC plans to develop a field near the Khafji oil field. To
spearhead the natural gas plan, Saudi Arabia formed the Saudi Gas Initiative, which will oversea the development of natural gas production in
conjunction with foreign oil companies. Overall, the natural gas plan has
been Saudi Arabia's attempt to diversify its economy away from crude oil
production, because export oil revenues currently make up almost 80 percent of the government's total earnings.

FURTHER READING

For an excellent overview of Saudi Arabia's history, see James Wynbrandt's *A
Brief History of Saudi Arabia* (New York: Checkmark Books, 2004) and Madawi
Al-Rasheed's *A History of Saudi Arabia* (New York: Cambridge University Press,
2002). Several works written during the oil shocks discuss Saudi Arabia's oil industry at that time. A few worthwhile texts include Peter Hobday's *Saudi Arabia
Today: An Introduction to the Richest Oil Power* (New York: St. Martin's Press,
1978) and William B. Quant's *Saudi Arabia in the 1980s: Foreign Policy, Security,
and Oil* (Washington, DC: Brookings Institution, 1981). For a good history of
Aramco, see Saudi Aramco's *Aramco and Its World* (Dhahran, Saudi Arabia: Saudi
Arabian Oil Company, 1995). For a discussion on Saudi Arabia's role in the for-
mation of OPEC, see Ali D. Johany's *The Myth of the OPEC Cartel: The Role of
Saudi Arabia* (New York: Wiley, 1980). For an excellent introduction to Islam, see
Thomas W. Lippman's *Understanding Islam: An Introduction to the Muslim World*
(New York: New American Library, 1982).

Chapter 16

Venezuela

VENEZUELA	
Official Name: Bolivarian Republic of Venezuela	
Location: Northern Caribbean coast of South America	
Capital: Caracas	
Major Port City: Caracas	
Independence from Colonial Rule: July 5, 1811, from Spain	
Population: 25 million	
National Currency: Bolívar (B)	
Official Language: Spanish	
Government: President Hugo Rafael Chávez Frías (since 1999)	
OPEC Member: Since 1960	
Proven Crude Oil Reserves: 78 billion bbl	
Proven Natural Gas Reserves: 148 Tcf	
National Oil Company: Petróleos de Venezuela S.A. (PdVSA)	
Crude Oil Production: 2.6 million bpd	
Natural Gas Production: 1 Tcf per year	
Production Ranking: Tenth largest	
Export Ranking: Fifth largest	

Map 16.1
Venezuela

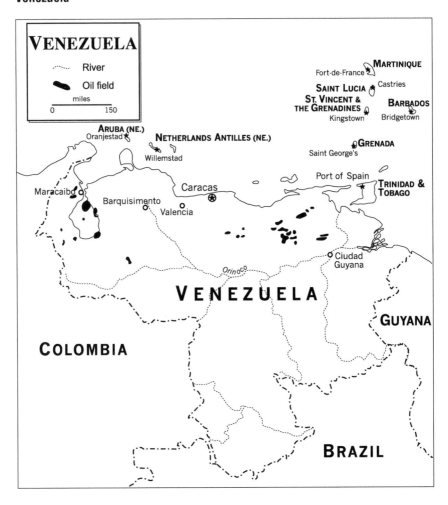

Venezuela is South America's largest petroleum exporter and the second largest source of U.S. crude oil imports. Within the past few years, Venezuela has faced political instability that has had a tremendous impact on its oil industry. Since the discovery of oil, the country's politics and petroleum industry have always been intimately connected. Expressions of nationalism in Venezuela have played most powerfully in decisions regarding the oil industry, such as the decision to nationalize, the formation of OPEC, and the most recent oil workers' strikes, which virtually shut down the country's production.

Located on the northern Caribbean coast of South America, Venezuela borders Colombia to the west, Guyana to the east, and Brazil to the south. Its northern coastlines touch the Caribbean Sea and the Atlantic Ocean. Much of Venezuela is tropical, with hot and humid temperatures. Venezuela is home to Lake Maracaibo, South America's largest lake (and is also considered one of the continent's most polluted body of water due to oil production) as well as Salto Angel (Angel Falls), which is the third largest waterfall in the world. The country is administered as a federal republic, with twenty-one states and eleven Caribbean dependencies. The national language is Spanish, but more than twenty-five other languages are spoken. This is because of Venezuela's diverse population, which includes a rich mixture of blacks, mestizos, mulattoes, and Amerindians.

Venezuela developed as one of the three countries that emerged from the collapse of Gran Colombia in 1830 (the others being Colombia and Ecuador). Almost all of South America had been under Spanish rule beginning in the sixteenth century. After the defeat of the Spanish in the northern region of the continent by Simón Bolívar, Venezuela, Colombia, and Ecuador were briefly united into one republic in 1821. Venezuela left the union in 1829 and faced a rocky beginning as an independent nation. Ruthless authoritarian leadership and social unrest plagued Venezuela until the mid-1900s. Amid revolutionary movements and political groups that rose and fell from power, the Acción Democrática (AD) remained a dominant political party in Venezuela between the 1940s and 1970s. AD began as a legal political party in 1941 (with roots reaching back to a late 1920s student movement). AD came to power in 1945 until a military coup in 1948, when it was banned. In 1958 elections took place, with AD becoming the ruling party from 1959 to 1969. The group's success lay in its ability to organize rural farmers and industrial workers as well as middle-class professionals. By 1948, the party membership grew to nearly half a million. AD held nationalization of the oil industry as a necessary, but long-term goal. AD remained influential in politics during the 1970s and 1980s through victorious presidential and congressional elections. In the 1980s, Venezuela faced high unemployment, inflation, and a large amount of international debt, which precipitated an economic crisis. In the late 1980s Venezuelan hostility toward the economic situation and growing corruption among leaders prompted a rash of strikes, protests, and two failed military coups in 1992. One of the coup leaders, Hugo Chávez, became president through popular election in 1998 and, again, in 2000. In 1999 he introduced a new constitution, which emphasized revolutionary social policies.

In December 2002, opponents of Chávez called for a general strike to demand that he resign or hold early elections. By December 2003 the opposition had collected 3.4 million signatures in a second petition demanding a referendum on Chávez's rule. In August 2004, Venezuelans voted on whether Chávez should finish term or not. Voting results for the referendum favored Chávez, making his opponents accuse him of fraud.[1] In the mean time, Venezuela's economy has suffered. Among other financial hardships, Venezuela is heavily dependent on its petroleum and natural resources.

VENEZUELA'S OIL INDUSTRY

Discovery

The first attempt to produce commercial quantities of oil in Venezuela began in the state of Táchira in 1878. Don Manuel Antonio Pulido received a concession to explore the region, which marked the first oil concession commercially exploited in Venezuela. His company, Compañía Petrolia de Táchira, was the first petroleum company in Venezuela. In 1914 the Caribbean Petroleum Company discovered the first oil field in Mené Grande, southeast of Lake Maracaibo. The company included several geologists and surveyors from Venezuela, Europe, and the United States. Running out of money, the company sold 51 percent to Shell. In 1917 the first oil refinery was built at San Lorenzo and the first shipment of oil was exported from there by Caribbean Petroleum. After World War I, a flood of oil companies arrived in Venezuela and made several major discoveries. The Venezuela Oil Concessions company, which Shell controlled, discovered La Rosa field, on the eastern shores of Lake Maracaibo, in 1922. Shortly after, several oil companies, including Standard Oil, discovered fields east of Lake Maracaibo. This area became known as the Bolívar coast fields and is still Venezuela's most important oil region. By 1928, Venezuela had become a full-fledged oil-producing country. Further discoveries within the country elevated it to one of the largest oil producers in the world.[2]

In addition to fields near Lake Maracaibo (a freshwater lake about 100 miles long and 60 miles wide), two more major regions were developed. In the 1920s oil companies began to expand into the Orinoco region—named after the Orinoco River, which it surrounds—which today is known as the Orinoco Heavy Oil Belt. This area is known for its heavy-oil deposits, which initially were not commercially attractive. In 1920 the

Creole Petroleum Corporation formed in Delaware to produce fields on Lake Maracaibo and was financed through the sale of stocks in New York. When Creole discovered commercial quantities of oil, SONJ bought the majority of the company in 1928, which then became the Creole Petroleum Corporation with Standard Oil of Venezuela as an affiliate. The potential of the area was discovered by Standard Oil of Venezuela in 1928 when it found the Quiriquire field in Monagas. The second region, known as the Barinas-Apure basin, is located in the southeast portion of the country and touches the Colombian border just north of the Orinoco River. Exploration in this region began in 1925. After numerous dry holes were drilled, oil was found in 1965. Until the 1960s, almost all of Venezuela's oil industry was in the hands of the foreign oil companies.

The Creole Petroleum Corporation held the largest concession areas, with Royal Dutch/Shell second. Foreign dominance in Venezuela, however, did not last, as nationalism spread through oil-producing countries and had a particularly profound impact on Venezuela, which nationalized its oil industry in the 1970s. In addition to onshore activities, Venezuela's oil industry developed offshore. In the 1930s, offshore activity began in Lake Maracaibo. In 1994, a subsidiary of PDVSA utilized modern drilling technology and boosted production in the already mature area of Maracaibo.

Industry Today

Since its discovery, Venezuela's oil has been in high demand within the United States because of its abundance as well as its close proximity to the United States. Venezuela is the second largest oil producer in the world after Saudi Arabia and was one of the founding members of OPEC (see Chapter 4). Venezuela's oil is heavy in consistency by international standards, and it needs to be improved in upgrading facilities that lighten the heavy crude and increase its value. Currently about 35 percent of Venezuela's oil production is consumed locally, leaving a substantial amount for export.

Venezuela has recoverable oil that is estimated to be about 78 billion bbl. Currently Venezuela's fields produce 2.6 million bpd, but the government plans an increase to 5.1 billion bpd by 2008. Venezuela has three major oil regions: Maracaibo-Falcon, Barinas-Apure, and the Orinoco Belt, which includes the large oil fields of Cerro Negro, Machete, Hamaca, and Zuata. Fields in southern Venezuela east of the Orinoco River are beginning to be explored near the Gulf of Paria. The majority of Venezuela's oil comes from the Orinoco Belt. Venezuela's fields are owned and operated through a system of concessions and joint ventures, primarily involving Venezuela's national oil company.

Most of Venezuela's oil is produced by a joint venture between major oil companies and the government-owned PDVSA. In 1975–1976, Venezuela nationalized its oil and natural gas industry and put it under the control of PDVSA. Today PDVSA is the ninth largest oil company in the world.[3] In 2003, PDVSA was split into two regional operating units: eastern and western. By doing this, the Venezuelan government hoped to redistribute managerial responsibilities to make the industry run efficiently. PDVSA holds at least a 50 percent share in the activities of foreign oil companies. Currently, there are six major joint venture companies working in Venezuela (ConocoPhillips, ExxonMobil, TotalFinaElf, Statoil, and ChevronTexaco), with the largest share held by ExxonMobil. PDVSA operates almost half the country's total production. Venezuela opened five offshore blocks on the Deltana Platform near the Venezuela-Trinidad maritime border in 2003. After bidding rounds, PDVSA took one, ChevronTexaco took two, and ConocoPhilips and Norway's Statoil each received a block.

Venezuela has recently suffered a major blow to its oil production due to severe political instability. Ongoing tension between Chávez and his opposition erupted in 2001 with the passing of laws that were designed to help the nation's poor. Many of the reforms were directed at Venezuela's oil industry and land rights. One year later he appointed a new board of directors to PDVSA, which prompted a general strike that lasted almost two months. During that time, output fell from 3.2 million bpd to below 40,000 bpd. In response, the government fired nearly half of its thousands of PDVSA workers. As a result, Venezuela's oil industry suffered a substantial loss of production and oil revenue.

Aside from the recent disruptions, Venezuela runs one of the world's largest refining systems. Domestic capacity stands at about 1.3 million bpd, with significant holdings in Curaçao, Europe (Germany, Sweden, Belgium, and the United Kingdom), and the United States (Lake Charles, LA, Corpus Christi, TX, and Houston, TX). More than one-third of Venezuela's refined products go to the United States and are distributed by Tulsa-based Citgo, which is PDVSA's U.S. refining and marketing subsidiary and one of the largest U.S. gasoline retailers.

Within Venezuela, PDVSA operates several refineries, which include El Palito, Puerto Cabello, Puerto de la Cruz, San Roque, and Anzoategui. The Paraguana Refining Center is currently the world's largest refinery; it includes several facilities connected by pipelines on the Paraguana peninsula. In addition to the refineries PDVSA operates within the country, it also supplies crude oil to refineries in which the company claims partial ownership. Hovensa, on the island of St. Croix, is owned by PDVSA and Amerada Hess Corporation. On the island of Curaçao, the Isla refinery is operated by PDVSA. Most of the oil processed at these refineries is reexported, with the majority going to the United States.

Because of the heavy characteristic of Venezuela's oil, it has to be upgraded before it can even be sent as crude oil to the refinery or to an importing country. Venezuela's industry has focused on new ways to utilize the extra heavy oil and bitumen it produces. The development of *orimulsion*, a branded product used by power plants as a boiler fuel, has recently expanded Venezuela's oil industry. It is named this because it comes from the Orinoco Belt. This product is imported by such countries as Canada, South Korea, and Italy. Bitor, a PDVSA subsidiary, manages the upstream and downstream aspects of it.

Venezuela has proven natural gas reserves of almost 150 Tcf and produces 1 Tcf per year of natural gas. Most of Venezuela's natural gas is associated, with a major portion of it reinjected or flared. Venezuelan gas reserves are located primarily offshore, near Trinidad and Tobago. While PDVSA has dominated the natural gas industry, it invited foreign companies in to explore for non-associated gas in 2001. Venezuela opened up offshore natural gas fields known as the Deltana Platform, a natural gas zone in Venezuela's Caribbean Sea, where there are 70 Bcf of natural gas leased to foreign oil companies such as ChevronTexaco for development.[4] Plans have been in the works for an LNG facility on the Paria Peninsula, which is expected to go onstream in 2008. Also, the construction of the Central-Occidental Interconnection pipeline, connecting natural gas fields to the Paraguana refinery, began at the end of 2004. Much of Venezuela's plans for expansion of its industry have been in the natural gas sector. Domestic consumption of gas has been low, but Venezuela hopes to improve it.

Venezuela has both gained and suffered because of its oil industry. Political and economic instability, however, have not dampened hopes for Venezuela to remain one of the world's largest oil producers. Oil income has resolved many economic and social problems as the government was able to create new industries and invest in public works. On the other hand, oil wealth also exacerbated existing social and political tensions. Allegations of mismanagement of oil funds by the government have become all too common.

FURTHER READING

A wealth of excellent books that examine the history of Venezuela exist in Spanish, but one in English has yet to be written. There are excellent texts, however, that examine Venezuela within the context of Latin American history. Within the eight volume series *The Cambridge History of Latin America* (Cambridge: Cambridge University Press, 1984–1995) are several good chapters that focus on Venezuela, such as Malcolm Deas's "Venezuela, Colombia, and Ecuador: the First Half-Century of Independence," vol. 3, 507–38, and "Columbia, Ecuador, and Venezuela: c. 1880–1930," vol. 5, 641–82, as well as Judith Ewell's chapter entitled "Venezuela since 1930," vol. 8, 727–90. For an overview of Venezuela's history and culture, see Mark Dinneen's *Culture and Customs of Venezuela* (Westport, CT: Greenwood Press, 2001). On the arrival of foreign oil companies in Venezuela, see B. S. McBeth's *Juan Vicente Gómez and the Oil Companies in Venezuela, 1908–1935* (New York: Cambridge University Press, 1983). For a thorough look at the nationalization of Venezuela's petroleum industry, see Laura Randall's *The Political Economy of Venezuelan Oil* (New York: Praeger, 1987) and Gustavo Coronel's *The Nationalization of the Venezuelan Oil Industry* (Lexington, MA: Lexington Books, 1983). See also Anibal R. Martinez's *Chronology of Venezuelan Oil* (London: Allen and Unwin, 1969) for a complete list of events related to oil up to 1967.

Notes

CHAPTER 1

1. Paul Roberts, *End of Oil on the Edge of a Perilous New World* (New York: Houghton Mifflin, 2004), 6.

2. Kate Van Dyke, *Fundamentals of the Petroleum Industry*, 4th ed. (Austin: University of Texas at Austin, 1997), 1.

3. George W. Grayson, "Oil and U.S.-Mexican Relations," *Journal of Interamerican Studies and World Affairs* 21, no. 4 (November 1979), 437.

4. Kenneth S. Deffeyes, *Hubbert's Peak: The Impending World Oil Shortage* (Princeton, NJ: Princeton University Press, 2001), 7, 10.

5. Jeff Gerth and Stephen Labaton, "Shell Withheld Reserves Data to Aid Nigeria," *New York Times*, March 19, 2004, 1.

6. Marie E. Williams, "Choices in Oil Refining: The Case of BP 1900–1960," *Business History* 26, no. 3 (November 1984), 308.

7. Steven A. Schneider, *The Oil Price Revolution* (Baltimore: Johns Hopkins University Press, 1983), 39.

8. Van Dyke, 327.

CHAPTER 2

1. Jules Abels, *The Rockefeller Billions* (New York: Macmillan, 1965), 1.

2. "Global 500: World's Largest Corporations," *Fortune*, July 26, 2004, 163.

3. Ernest Smith, John S. Dzienkauski, Owen L. Anderson, Gary B. Conine, John S. Love, *International Petroleum Transactions* (Denver: Rocky Mountain Mineral Law Foundation, 1993), 25.

4. Portions of chart from Angela Clarke, *Bahrain Oil and Development, 1929–1989* (London: Immel, 1991), 49.

5. The name first emerges in 1952 because of a U.S. government investigation into the possibility of the companies forming a global oil cartel. See Anthony Sampson, *The Seven Sisters: The Great Oil Companies and the World They Shaped* (New York: Viking Press, 1975), 59.

6. Sampson, 43–44.

7. E. B. Brossard, *Petroleum: Politics and Power* (Tulsa, OK: PennWell Books, 1983), 12.

8. Sampson, 143–147.

9. Sidney Swensrud, *Gulf Oil: The First Fifty Years, 1901–1951* (New York: Newcomen Society in North America, 1951), 7–10.

10. Craig Thompson, *Since Spindletop: A Human Story Of Gulf's First Half Century* (Pittsburgh: Gulf Oil, 1950), 14, 16.

11. Charles R. Dechert, *Ente Nazionale Idrocarburi* (Leiden: E. J. Brill, 1963), 5.

12. Smith, 303.

13. Henrietta M. Larson, Evelyn H. Knowlton, and Charles S. Popple, *History of Standard Oil Company (New Jersey)*, vol. 3 (New York: Harper and Brothers, 1955–1971); *New Horizons, 1927–1950* (New York: Harper and Row, 1971), 308–14.

CHAPTER 3

1. Lázaro Cárdenas, "The President's Message to the Mexican Nation," in *Messages to the Mexican Nation on the Oil Question* (Mexico, no pub, 1938), 5.

2. Laura Randall, *The Political Economy of Venezuelan Oil* (New York: Praeger, 1987), 210.

3. Ibid., 93–94.

4. Ernest Smith, John S. Dzienkowski, Owen L. Anderson, Gary B. Conine, John S. Love, *International Petroleum Transactions* (Denver: Rocky Mountain Mineral Law Foundation, 1993), 332.

5. Jonathan Brown, *Oil and Revolution in Mexico* (Berkeley: University of California Press, 1993), 311.

6. Herbert Klein, "American Oil Companies in Latin America: The Bolivian Experience," *Inter-American Economic Affairs* 18, no. 2 (Autumn 1964): 55; George Philip, *Oil and Politics in Latin America* (Cambridge: Cambridge University Press, 1982), 47.

7. Cárdenas, "The President's Message," 5–12.

8. Gustave Coronel, *The Nationalization of the Venezuelan Oil Industry* (Lexington, MA: Lexington Books, 1983), 23–33.

9. Ernest Smith, John S. Dzienkowski, Owen L. Anderson, Gary B. Conine, and John S. Lowe, *International Petroleum Transactions* (Denver, CO: Rocky Mountain Mineral Law Foundation, 1993), 305.

10. Anthony Harrup, "Chinese Oil Companies Bid for Mexican Natural Gas Block," *Wall Street Journal*, November 18, 2003, www.wsj.com.

11. *Wall Street Journal,* "Venezuela to Offer 2 Gas Blocks by End of Nov," November 20, 2003. http://www.wsj.com.

12. James Bamberg, *British Petroleum and Global Oil, 1950–1975: The Challenge of Nationalism* (Cambridge: Cambridge University Press, 2000), 68.

13. Michael Brown, "The Nationalization of the Iraqi Oil Company," *International Journal of Middle Eastern Studies* 10, no. 1 (February 1979), 121.

CHAPTER 4

1. Yamani commenting on the first oil shock in Jeffery Robinson, *Yamani: The Inside Story* (London: Simon and Schuster, 1988), 82.

2. Francisco Parra, *Oil Politics* (New York: I. B. Tauris, 2004), 91, 94.

3. Ibid., 93–94.

4. "Russia Criticizes OPEC," *Weekly Petroleum Argus* 33, no. 41, October 20, 2003, 8; "Russia, Saudi Arabia Sign Oil Pact," *Wall Street Journal Europe,* September 3, 2003, 1.

5. For more on the Israeli-Palestinian conflict, see Charles D. Smith's *Palestine and the Arab-Israeli Conflict* (New York: St. Martin's Press, 1988).

6. *Time,* various issues from October 1973 to May 1974.

7. The Arab League formed in 1945 and serves as a regional organization of Arab states. The focus of the organization is to promote political, economic, and cultural interests of the member states.

8. Miriam Camps, *"First World" Relationships: The Role of the OECD* (New York: Council of Foreign Relations, 1975), 5.

9. Camps, 33.

CHAPTER 5

1. Rilwanu Lukman, *OPEC Bulletin* (May 1998), 4.

2. Markos Mamalakis, "The New International Economic Order: Centerpiece Venezuela," *Journal of Interamerican Studies and World Affairs* 20, no. 3 (August 1978): 283.

3. "Venezuela, Zimbabwe Sign Energy Cooperation Pact," *Wall Street Journal* (February 26, 2004), www.wsj.com; for more on Zimbabwe, see Oyekan Owomoyela, *Culture and Customs of Zimbabwe* (Westport, CT: Greenwood Press, 2002).

4. Laura Randall, *The Political Economy of Mexican Oil* (New York: Praeger, 1989), 185.

5. "Suit Seeks to Block Drilling in Disputed Venezuela Area," *Wall Street Journal* (February 25, 2004), www.wsj.com.

6. Gregg Hitt, Neil King Jr., and Jess Bravin, "Bush Will Allow Canada to Bid for Iraq Contracts," *Wall Street Journal,* January 14, 2004, www.wsj.com.

CHAPTER 6

1. Greenpeace advertisement.

2. Michael Klare, *Resource Wars: The New Landscape of Global Conflict* (New York: Metropolitan Books, 2001), 15–16.

3. Peter Colley, *Reforming Energy: Sustainable Futures and Global Labour* (London: Pluto Press, 1997), 44–45.

4. Hart Publications, *Fifty Years of Offshore Oil and Gas Development* (Houston, TX: Hart Publications, 1997), 95.

5. Muhammad Sadiq and John C. McCain, *The Gulf War Aftermath: An Environmental Tragedy* (Boston: Kluwer Academic, 1993), 2–3; Committee on Oil in the Sea National Research Council, *Oil in the Sea III: Inputs, Fates, and Effects?* (Washington, DC: National Academies Press, 2003), 135.

6. Rory Cox, "Russian Big Oil Redraws Pipe Dream," *Washington Free Press* 62 (March 2003), www.washingtonfreepress.org/62/russianBigOil.html; Sabrina Tavernise, "Oil Rush in Siberia Puts Other Treasures at Risk," *New York Times*, October 5, 2003, 3.

7. James A. Fay and Dan S. Golomb, *Energy and the Environment* (New York: Oxford University Press, 2002), 258.

8. Alvaro Silva-Calderón, "OPEC Statement to the Conference of the Parties" (speech, Ninth Conference of the Parties to the UN Framework Convention on Climate Change, Milan, Italy, December 1–12, 2003), www.opec.org/NewsInfo/Speeches.

9. Simon Collins, "Oil Giant in Environmental Gun," *New Zealand Herald*, February 18, 2004, www.nzherald.co.nz.

10. Energy Information Administration, "International Energy Outlook, World Oil Markets," April 2004, www.eia.doe/gov/oiaf/ieo/oil.html.

11. "Petróleos Mexicanos (Pemex), Mexico's State-Owned Oil Company," *Latin Trade* 12, no. 3 (March 2003), 14.

12. United Nations Environment Programmes, *Understanding Climate Change: A Beginners Guide to the UN Framework Convention to Its Kyoto Protocol* (Châtelaine, Switz.: United Nations, 2002).

13. See Chapter 8 for more information on the World Bank Group.

CHAPTER 7

1. Ken Saro-Wiwa's last words before he was hanged, *Daily Times* (Nigeria), November 13, 1995, 1.

2. Myrna Santiago, "Rejecting Progress in Paradise: Huastecs, the Environment, and the Oil Industry in Veracruz, Mexico, 1900–1935," *Environmental History* 3, no. 2 (1998), 169–88.

3. Gustavo Coronel, *The Nationalization of the Venezuelan Oil Industry* (Lexington, MA: Lexington Books, 1983), 11.

4. Jedrzej Georg Frynas, *Oil in Nigeria: Conflict and Litigation Between Oil Companies and Village Communities* (Berlin: Lit Verlag, 2000), 46.

5. Frynas, 95.

6. James Bamberg, *British Petroleum and Global Oil, 1950–1975: The Challenge of Nationalism* (Cambridge: Cambridge University Press, 2000), 41.

7. Bill Vann, "US, British Air Strikes Kill Iraqi Oil Workers," *Wall Street Journal*, December 3, 2002, http://www.wsj.com.

8. Øystein Noreng, *Oil and Islam: Social and Economic Issues* (New York: Wiley, 1997), 5–6.

9. David Luhnow and José de Códoba, "Academics' Study Backs Fraud Claim in Chávez Election," *Wall Street Journal*, September 7, 2004, 18.

10. Laura Randall, *The Political Economy of Venezuelan Oil* (New York: Praeger, 1987), 6–7.

11. Mark Long, "Nigeria's Senate Wants to Fine Shell $1.5 Billion," *Wall Street Journal*, August 26, 2004, A2.

CHAPTER 8

1. Markos Mamalakis, "The New International Economic Order: Centerpiece Venezuela," *Journal of Interamerican Studies and World Affairs* 20, no. 3 (August 1978): 283.

2. *Sustainable* meaning that a country can meet its population's need in the present without compromising future generations' needs.

3. See Chapter 3 for more on the first oil shock.

4. Øystein Noreng, *Oil and Islam: Social and Economic Issues* (New York: Wiley, 1997), 135, 130–140; Mamalakis, 288.

5. James H. Street, "Coping with Energy Shocks in Latin America: Three Responses," *Latin American Research Review* 17, no. 3 (1982), 136–37.

6. Alan Gelb and Associates, *Oil Windfalls: Blessing or Curse?* (Oxford: Oxford University Press, 1988), 241.

7. Abbas Alnasrawi, *The Economy of Iraq: Oil, Wars, Destruction of Development and Prospects, 1950–2010* (Westport, CT: Greenwood Press, 1994), 71.

8. Sheikh Mohammed Ab-Alkhail, "OPEC Aid: A Question of Solidarity," in *OPEC AID and the Challenge of Development*, ed. Abdelkader Benamara and Sam Ifeagwu (London: Croom Helm, 1987), 10.

9. Edgardo Lander, "The Impact of Neoliberal Adjustment in Venezuela, 1989–1993," *Latin American Perspectives* 23, no. 3 (Summer 1996): 50–56.

10. Manuel Pastor, "Latin America, The Debt Crisis, and the International Monetary Fund," *Latin American Perspectives* 16, no. 1 (Winter 1989), 91–92.

11. Editorial, "Rescind This Decision," *Daily Standard* (Nigeria) October 5, 1978, 3.

12. Laura Randall, *The Political Economy of Venezuelan Oil* (New York: Praeger, 1987), 197–198.

13. Energy Information Administration, "Country Analysis Briefs, Norway," December 2004, www.eia.doe.gov/emeu/cabs/norway.html.

14. Marshall I. Goldman, *The Piratization of Russia* (New York: Routledge, 2003), 148.

15. "Roman Abroamovich, Head of Russian Oil Company Sibneft, Called Off His Company's Planned Merger with Yukos," *Corporate Financing Week*, December 15, 2003, 7.

CHAPTER 9

1. Thomas Gold in Oliver Morton's "Fuel's Paradise," *Wired*, July 2000, 170.

2. "Nine Billion People by 2050," *BBC Online*, February 28, 2001, http://news.bbc.co.uk/1/world/1194030.html.

3. Maggie Shiels, "Environment Drives Hummer vs Hybrid Row," *BBC Online*, June 1, 2004, http://news.bbc.co.uk/1/hi/business/3749377.html.

4. Energy Information Administration, "International Energy Outlook 2004, World Oil Markets" April 2004, www.eia.doe.gov/oiaf/ieo/oil.html.

5. Greg Ip and Bhushan Bahree, "High Demand for Oil Could Mean High Prices are Here for Long Run," *Wall Street Journal*, August 20, 2004, 1.

6. Energy Information Administration, "International Energy Outlook, Natural Gas," 2004, www.eia.doe.gov/oiaf/ieo/nat_gas.html.

7. *OPEC Annual Statistical Bulletin 2003* (Vienna: OPEC, 2004), 10.

8. Øystein Noreng, *Oil and Islam: Social and Economic Issues* (New York: Wiley, 1997), 119–120.

9. Norges Bank, *Petroleum Fund* (2005), www.norgesbank.no/english/petro leum_fund/html.

10. Morton, 160–172.

11. "How OPEC's Fear of $5 Oil Led to $50 Oil," *Economist*, August 21, 2004, 59.

12. "Awash in a Gusher of Cash, Oil Firms are Reluctant Investors," *Wall Street Journal*, August 26, 2004, 1, 2.

CHAPTER 11

1. Robert Miller, *Mexico: A History* (Norman: University of Oklahoma Press, 1985), 283–308.

2. "Global 500: World's Largest Corporations," *Fortune*, July 26, 2004, 163.

3. Tim Weiner, "Corruption and Waste Bleed Mexico's Corrupt Oil Lifeline," *New York Times*, January 21, 2003, 1.

CHAPTER 12

1. Sola Odunfa, "Anti-Corruption Lessons for Nigeria," *BBCOnline*, June 22, 2004, news.bbc.co.uk.

2. J.K. Onoh, *The Nigerian Oil Economy: From Prosperity to Glut* (New York: St. Martin's Press, 1983), 43.

CHAPTER 13

1. Thomas K. Derry, *A History of Modern Norway, 1814–1972* (Oxford: Clarendon Press, 1973).

2. Brent F. Nelsen, *The State Offshore: Petroleum, Politics, and State Intervention on the British and Norwegian Continental Shelves* (Westport, CT: Praeger, 1991), 15.

3. Hart Publications, *Fifty Years of Offshore Oil and Gas Development* (Houston, TX: Hart Publications, 1997), 10–11, 15.

4. Energy Information Administration, "Country Analysis Briefs, Norway," December 2004, www.eia.doe.gov/emeu/cabs/norway.html; "Global 500: World's Largest Corporations," *Fortune*, July 26, 2004, 163.

CHAPTER 14

1. Marshall I. Goldman, *The Piratization of Russia* (New York: Routledge, 2003), 13.

2. E. B. Brossard, *Petroleum: Politics and Power* (Tulsa, OK: PennWell Books, 1983), 13.

3. George S. Gibb and Evelyn H. Knowlton, *History of Standard Oil Company (New Jersey)*, vol. 2 (New York: Harper and Brothers, 1955–1971); *The Resurgent Years, 1911–1927* (New York: Harper and Brothers, 1956), 329.

4. F. C. Gerretson, *The History of Royal Dutch* (The Hague: Royal Dutch Petroleum Company, 1955), 157.

5. Gibb and Knowlton, *History of Standard Oil Company (New Jersey)*, vol. 2, *The Resurgent Years*, 330.

6. Bennett H. Wall, *Growth in a Changing Environment: A History of Standard Oil Company (New Jersey)*, (New York: McGraw-Hill, 1988), 332.

7. "Russia Criticizes OPEC," *Weekly Petroleum Argus*, October 20, 2003, 8, www.argusonline.com.

8. Goldman, *The Piratization of Russia*, 117.

9. Conrad De Aenlle, "Russia's Oil Industry, Caught in a Tug of War," *New York Times*, November 9, 2003, 4.

10. Energy Information Administration, "Country Analysis Briefs, Sakhalin Fact Sheet," January 2005, www.eia.doe.gov/emeu/cabs/sakhalin.html.

11. Energy Information Administration, "Country Analysis Briefs, Russia," February 2005, www.eia.doe.gov/emeu/cabs/russia.html.

CHAPTER 15

1. International Energy Agency, *Middle East Oil and Gas* (Paris: OECD/IEA, 1995), 188.

CHAPTER 16

1. David Luhnow and José de Córdoba, "Academics' Study Backs Fraud Claim in Chávez Election," *Wall Street Journal*, September 7, 2004, 18.

2. Bernard Mommer, *Global Oil and the Nation State* (Oxford: Oxford University Press, 2002), 109–120.

3. "Global 500: World's Largest Corporations," *Fortune*, July 26, 2004, 163.

4. "Chávez Inauguró Explotación de Gas de la Chevron Texaco en Plataforma Deltana" [Chávez Unvailed ChevronTexaco's Gas Production on the Deltana Platform], *El Nacional*, August 7, 2004, www.el-nacional.com.

Bibliography

Abels, Jules. *The Rockefeller Billions*. New York: Macmillan, 1965.

Ahrari, Mohammed E. *OPEC: The Failing Giant*. Lexington: University Press of Kentucky, 1986.

Alnasrawi, Abbas. *The Economy of Iraq: Oil, Wars, Destruction of Development and Prospects, 1950–2010*. Westport, CT: Greenwood Press, 1994.

———. *Iraq's Burdens: Oil, Sanctions, and Underdevelopment*. Westport, CT: Greenwood Press, 2002.

Amnesty International. *Nigeria, Repression of Women's Protests in Oil-Producing Delta Region*. London: International Secretariat, 2003.

Amuzegar, Jahangir. *Managing the Oil Wealth: OPEC's Windfalls and Pitfalls*. New York: I. B. Tauris, 2001.

Anderson, Robert. *Fundamentals of the Petroleum Industry*. Norman: University of Oklahoma Press, 1984.

Arnove, Anthony, ed. *Iraq Under Siege: The Deadly Impact of Sanctions and War*. Cambridge, MA: South End Press, 2002.

Bacher, John. *Petrotyranny*. Toronto: Dunburn Press, 2000.

Baer, M. Delal, and Sidney Weintraub. *The NAFTA Debate: Grappling with Unconventional Trade*. Boulder, CO: Lynne Rienner, 1994.

Bamberg, James. *British Petroleum and Global Oil, 1950–1975: The Challenge of Nationalism*. Cambridge: Cambridge University Press, 2000.

Bartley, Kim, and Donnacha O' Brian. *This Revolution Will Not Be Televised*. Dublin: Irish Film Board, 2002.

Bartsch, Ulrich, and Benito Muller. *Fossil Fuels in a Changing Climate: Impacts of the Kyoto Protocol and Developing Country Participation*. Oxford: Oxford University Press, 2000.

Benamara, Abdelkader, and Sam Ifeagwu, eds. *OPEC AID and the Challenge of Development*. London: Croom Helm, 1987.

Bennis, Phyliss, and Michel Moushabeck. *Beyond the Storm: A Gulf Crisis Reader*. Northampton, MA: Interlink, 1998.

Bermúdez, Antonio J. *The Mexican National Petroleum Industry: A Case Study in Nationalization.* San Jose, CA: Stanford University Press, 1963.

Bethell, Leslie, ed. *The Cambridge History of Latin America.* Cambridge: Cambridge University Press, 1984.

Brossard, E. B. *Petroleum: Politics and Power.* Tulsa, OK: PennWell Books, 1983.

Brown, Jonathan. *Oil and Revolution in Mexico.* Berkeley: University of California Press, 1993.

Brown, Michael. "The Nationalization of the Iraqi Petroleum Company." *International Journal of Middle Eastern Studies* 10, no. 1 (February 1979): 107–24.

Camp, Miriam. *"First World" Relationships: The Role of the OECD.* New York: Council of Foreign Relations, 1975.

Cárdenas, Lázaro. *Messages to the Mexican Nation on the Oil Question.* Mexico City, no pub., 1938.

Clarke, Angela. *Bahrain Oil and Development, 1929–1989.* London: Immel, 1991.

Cole, Ken. *Economy-Environment-Development-Knowledge.* New York: Routledge, 1999.

Colley, Peter. *Reforming Energy: Sustainable Futures and Global Labour.* London: Pluto Press, 1997.

Committee on Oil in the Sea National Research Council. *Oil in the Sea III: Inputs, Fates, and Effects?* Washington, DC: National Academies Press, 2003.

Conaway, Charles F. *The Petroleum Industry: A Nontechnical Guide.* Tulsa, OK: PennWell, 1999.

Coronel, Gustavo. *The Nationalization of the Venezuelan Oil Industry.* Lexington, MA: Lexington Books, 1983.

Dechert, Charles R. *Ente Nazionale Idrocarburi.* Leiden: E. J. Brill, 1963.

Deffeyes, Kenneth S. *Hubbert's Peak: The Impending World Oil Shortage.* Princeton, NJ: Princeton University Press, 2001.

Derry, Thomas K. *A History of Modern Norway, 1814–1972.* Oxford: Clarendon Press, 1973.

Dinneen, Mark. *Culture and Customs of Venezuela.* Westport, CT: Greenwood Press, 2001.

Dodge, Toby. *The Inventing of Iraq: The Failure of Nation Building and a History Denied.* New York: Columbia University Press, 2003.

Duke, Paul. *A History of Russia: c. 882–1996.* Durham, NC: Duke University Press, 1998.

Energy Information Administration. "International Energy Outlook 2004" (April 2004). www.eia.doe.gov/oiaf/ieo/world.html.

Falola, Toyin. *Culture and Customs of Nigeria.* Westport, CT: Greenwood Press, 2001.

———. *History of Nigeria.* Westport, CT: Greenwood Press, 1999.

Fay, James A., and Dan S. Golomb. *Energy and the Environment.* New York: Oxford University Press, 2002.

Ferrier, Ronald W. *The History of the British Petroleum Company*. Vols. 1 and 2. New York: Cambridge University Press, 1982.

Fisher, Sydney, and William Ochsenwald. *The Middle East: A History*. Vol. 2. 5th ed. New York: McGraw-Hill, 1990.

Frynas, Jedrzej Georg. *Oil in Nigeria: Conflict and Litigation Between Oil Companies and Village Communities*. Hamburg, Ger.: Lit Verlag, 2001.

Gaither, Roscoe B. *Expropriation in Mexico*. New York: William Morrow, 1940.

Gardner, Richard N. *Negotiating Survival: Four Priorities After Rio*. New York: Council on Foreign Relations, 1992.

Gelb, Alan, and Associates. *Oil Windfalls: Blessing or Curse?* Oxford: Oxford University Press, 1988.

Gerretson, F. C. *The History of Royal Dutch*. The Hague: Royal Dutch Petroleum Company, 1955.

Gibb, George S., and Evelyn H. Knowlton. *The History of Standard Oil Company (New Jersey)*, vol. 2, *The Resurgent Years, 1911–1927*. New York: Harper and Brothers, 1956.

"Global 500: World's Largest Corporations." *Fortune*, July 26, 2004, 163–80.

Goldman, Marshall. *The Piratization of Russia*. New York: Routledge, 2003.

Grayson, George W. "Oil and U.S.-Mexican Relations." *Journal of Interamerican Studies and World Affairs* 21, no. 4 (November 1979): 427–56.

Grubb, Michael. *The "Earth Summit" Agreements: A Guide and Assessment: An Analysis of the Rio '92 UN Conference on Environment and Development*. London: Earthscan Publications, 1993.

Hart Publications. *Fifty Years of Offshore Oil and Gas Development*. Houston, TX: Hart Publications, 1997.

Hassan, Hamdi A. *The Iraqi Invasion of Kuwait: Religion, Identity, and Otherness in the Analysis of War and Conflict*. London: Pluto Press, 1999.

Hell, Knut, ed. *The Cambridge History of Scandinavia*, Vol. 1, *Prehistory to 1520*. Cambridge: Cambridge University Press, 2003.

Herspring, Dale R. *Putin's Russia: Past Imperfect, Future Uncertain*. Lanham, MD: Rowman and Littlefield, 2003.

Hidy, Ralph W., and Muriel E. Hidy. *History of Standard Oil Company (New Jersey)*, vol. 1, *Pioneering in Big Business, 1882–1971*. New York: Harper & Brothers, 1955.

Hobday, Peter. *Saudi Arabia Today: An Introduction to the Richest Oil Power*. New York: St. Martin's Press, 1978.

Howarth, Stephen. *A Century of Oil: The "Shell" Transport and Trading Company, 1897–1997*. London: Weidenfeld and Nicolson, 1997.

Ikeh, Goddy. *Nigerian Oil Industry: The First Three Decades (1958–1988)*. Lagos, Nigeria: Starledger Communications, 1991.

Inati, Shams C., ed. *Iraq: Its History, People, and Politics*. Amherst, NY: Humanity Books, 2003.

International Energy Agency. *Middle East Oil and Gas*. Paris: OECD/IEA.

Johany, Ali D. *The Myth of the OPEC Cartel: The Role of Saudi Arabia.* New York: Wiley, 1980.

Khan, Sarah Ahmad. *Nigeria: The Political Economy of Oil.* New York: Oxford University Press, 1994.

Kirkwood, Burton. *History of Mexico.* Westport, CT: Greenwood Press, 2002.

Klare, Michael T. *Resource Wars: The New Landscape of Global Conflict.* New York: Metropolitan Books, 2001.

Klein, Herbert. "American Oil Companies in Latin America: The Bolivian Experience." *Inter-American Economic Affairs* 18, no. 2 (Autumn 1964): 47–72.

Lander, Edgardo. "The Impact of Neoliberal Adjustment in Venezuela, 1989–1993." *Latin American Perspectives* 23, no. 3 (Summer 1996): 50–73.

Lane, David, ed. *The Political Economy of Russian Oil.* New York: Rowman and Littlefield, 1999.

Lange, Nicholas de. *An Introduction to Judaism.* New York: Cambridge University Press, 2000.

Larson, Henrietta M., Evelyn H. Knowlton, and Charles S. Popple. *The History of Standard Oil Company (New Jersey).* Vol. 3, *New Horizons, 1927–1950.* New York: Harper and Row, 1971.

Lieven, Anatol. *Chechnya: Tombstone of Russian Power.* New Haven, CT: Yale University Press, 1998.

Lippman, Thomas W. *Understanding Islam: An Introduction to the Muslim World.* New York: New American Library, 1982.

Lynn, Stuart. *Economic Development: Theory and Practice for a Divided World.* Upper Saddle River, NJ: Prentice Hall, 2003.

Maachou, Abdelkader. *OAPEC: An International Organization for Economic Cooperation and an Instrument for Regional Integration.* New York: St. Martin's Press, 1983.

Mallakh, Ragaei El. *Petroleum and Economic Development: The Cases of Mexico and Norway.* Lexington, MA: Lexington Books, 1984.

Mamalakis, Markos. "The New International Economic Order: Centerpiece Venezuela." *Journal of Interamerican Studies and World Affairs* 20, no. 3 (August 1978): 265–95.

Manby, Browen. *The Price of Oil: Corporate Responsibility and Human Rights Violations in Nigeria's Oil Producing Communities.* New York: Human Rights Watch, 1999.

Martinez, Anibal R. *Chronology of Venezuelan Oil.* London: Allen and Unwin, 1969.

McBeth, B. S. *Juan Vicente Gómez and the Oil Companies in Venezuela, 1908–1935.* New York: Cambridge University Press, 1983.

Miller, Robert. *Mexico: A History.* Norman: University of Oklahoma Press, 1985.

Mommer, Bernard. *Global Oil and the Nation State.* Oxford: Oxford University Press, 2002.

Morton, Oliver. "Fuel's Paradise." *Wired,* July 2000, 160–72.

Nelsen, Brent F. *The State Offshore: Petroleum, Politics, and State Intervention on the British and Norwegian Continental Shelves.* New York: Praeger, 1991.

Noreng, Øystein. *Oil and Islam: Social and Economic Issues.* New York: Wiley, 1997.

———. *The Oil Industry and Government Strategy in the North Sea.* London: Croom Helm, 1980.

Nwosu, Nereus I. "The Politics of Oil Subsidy in Nigeria." *Africa* (Italy) (1996): 80–94.

Obi, Cyril. *Changing Forms of Identity Politics in Nigeria Under Economic Adjustment: The Case of the Oil Minorities Movement in the Niger Delta.* Uppsala, Swed.: Nordiska Afrikainstitutet, 2001.

Onoh, J.K. *The Nigerian Oil Economy: From Prosperity to Glut.* New York: St. Martin's Press, 1983.

Organization for Economic Cooperation and Development. *The OECD: History, Aims, Structure.* Paris: OECD, 1960.

Organization of the Petroleum Exporting Countries. *Basic Oil Industry Information.* Vienna: OPEC, 1983.

Osaghae, Eghosu. *Crippled Giant: Nigeria Since Independence.* Bloomington: Indiana University Press, 1998.

Owomoyela, Oyekan. *Culture and Customs of Zimbabwe.* Westport, CT: Greenwood Press, 2002.

Parra, Francisco. *Oil Politics.* New York: I. B. Tauris, 2004.

Pastor, Manuel. "Latin America, the Debt Crisis, and the International Monetary Fund." *Latin American Perspectives* 16, no. 1 (Winter 1989): 79–110.

Payer, Cheryl. *The World Bank: A Critical Analysis.* New York: Monthly Review Press, 1982.

Pelletière, Stephen C. *The Iran-Iraq War.* Westport, CT: Greenwood Press, 1992.

———. *Iraq and the International Oil System: Why America Went to War in the Gulf.* Westport, CT: Praeger, 2001.

Penrose, Edith. *The Large International Firm in Developing Countries: The International Petroleum Industry.* London: George Allen and Unwin, 1968.

"Petróleos Mexicanos, Pemex, Mexico's State-Owned Oil Company." *Latin Trade* 12, no. 3 (March 2003): 14.

Philip, George. *Oil and Politics in Latin America: Nationalist Movements and State Companies.* Cambridge: Cambridge University Press, 1982.

Quant, William B. *Saudi Arabia in the 1980s: Foreign Policy, Security, and Oil.* Washington, DC: Brookings Institution, 1981.

Ramamurti, Ravi. "The Impact of Privatization on the Latin American Debt Problem." *Journal of Interamerican Studies and World Affairs* 34, no. 2 (Summer 1992): 93–125.

Randall, Laura. *The Political Economy of Mexican Oil.* New York: Praeger, 1989.

———. *The Political Economy of Venezuelan Oil.* New York: Praeger, 1987.

Rasheed, Madawi Al-. *A History of Saudi Arabia.* New York: Cambridge University Press, 2002.

Riasanovsky, Nicholas V. *A History of Russia*. 6th ed. New York: Oxford University Press, 2000.

Roberts, Paul. *End of Oil on the Edge of a Perilous New World*. New York: Houghton Mifflin, 2004.

Robinson, Jeffery. *Yamani: The Inside Story*. London: Simon and Schuster, 1988.

Ruhani, Fu'ad. *A History of O.P.E.C.* New York: Praeger, 1971.

Sadiq, Muhammad, and John C. McCain. *The Gulf War Aftermath: An Environmental Tragedy*. Boston: Kluwer Academic, 1993.

Sampson, Anthony. *Seven Sisters: The Great Oil Companies and the World They Shaped*. New York: Viking Press, 1975.

Santiago, Myrna. "Rejecting Progress in Paradise: Huastecs, the Environment, and the Oil Industry in Veracruz, Mexico, 1900–1935." *Environmental History* 3, no. 2 (1998): 169–88.

Saro-Wiwa, Ken. *Genocide in Nigeria: The Ogoni Tragedy*. London: Saros International, 1992.

Saudi Aramco. *Aramco and Its World*. Dhahran, Saudi Arabia: Saudi Arabian Oil Company, 1995.

Schneider, Steven A. *The Oil Price Revolution*. Baltimore: Johns Hopkins University Press, 1983.

Schultze, Sydney. *Culture and Customs of Russia*. Westport, CT: Greenwood Press, 2000.

Service, Robert. *A History of Modern Russia: From Nicholas II to Putin*. London: Penguin Books, 2003.

Silva-Calderón, Alvaro. "OPEC Statement to the Conference of the Parties." Speech, Ninth Conference of the Parties to the UN Framework Convention on Climate Change, Milan, Italy, December 1–12, 2003. http://www.opec.org/NewsInfo/Speeches.

Smith, Charles D. *Palestine and the Arab-Israeli Conflict*. New York: St. Martin's Press, 1988.

Smith, Ernest, John S. Dzienkowski, Owen L. Anderson, Gary B. Conine, John S. Love. *International Petroleum Transactions*. Denver, CO: Rocky Mountain Mineral Law Foundation, 1993.

Smith, Peter H. "The Political Impact of Free Trade on Mexico." *Journal of Interamerican Studies and World Affairs* 34, no. 1 (Spring 1992): 1–25.

Standish, Peter, and Steven M. Bell. *Culture and Customs of Mexico*. Westport, CT: Greenwood Press, 2004.

Street, James H. "Coping with Energy Shocks in Latin America: Three Responses." *Latin American Research Review* 17, no. 3 (1982): 128–47.

Swensrud, Sidney. *Gulf Oil: The First Fifty Years, 1901–1951*. New York: Newcomen Society in North America, 1951.

Tanzer, Michael. *The Political Economy of International Oil and the Undeveloped Countries*. Boston: Beacon Press, 1969.

Thompson, Craig. *Since Spindletop: A Human Story of Gulf's First Half Century.* Pittsburgh, PA: Gulf Oil, 1950.

Tripp, Charles. *A History of Iraq.* Cambridge: Cambridge University Press, 2002.

Turner, Terisa, and Leigh S. Brownhill. *Why Women Are at War with Chevron: Nigerian Subsistence Struggles Against the International Oil Industry.* New York: International Working Group, 2003.

United Nations Environment Programmes. *Understanding Climate Change: A Beginners Guide to the UN Framework Convention to Its Kyoto Protocol.* Châtelaine, Switz.: United Nations, 2002.

Van Dyke, Kate. *Fundamentals of Petroleum.* Austin: University of Texas Press, 1997.

Velasco, Jesus Agustin. *Impacts of Mexican Oil Policy on Economic and Political Development.* Lexington, MA: Lexington Books, 1983.

Wall, Bennett H. *A History of Standard Oil Company (New Jersey),* vol. 4, *Growth in a Changing Environment.* New York: McGraw-Hill, 1988.

Weissman, Robert. "Why We Protest." *Washington Post,* September 10, 2001, 21.

Williams, Marie W. "Choices in Oil Refining: The Case of BP 1900–60." *Business History* 26, no. 3 (November 1984): 307–328.

Wirth, John D., ed. *Latin American Oil Companies and the Politics of Energy.* Lincoln: University of Nebraska Press, 1985.

World Trade Organization. *Understanding the WTO.* 4th ed. Geneva: WTO, 2003.

Wunder, Sven. *Oil Wealth and the Fate of the Forest.* London: Routledge, 2003.

Wynbrandt, James. *A Brief History of Saudi Arabia.* New York: Checkmark Books, 2004.

Yergin, Daniel. *The Prize: The Epic Quest for Oil, Money, and Power.* New York: Simon and Schuster, 1993.

Zarsky, Lyuba. *Human Rights and the Environment.* Sterling, VA: Earthscan, 2002.

Ziegler, Charles E. *History of Russia.* Westport, CT: Greenwood Press, 1999.

PERIODICALS

The Daily Standard (Nigeria)

The Daily Times (Nigeria)

The Economist

International Energy Agency Statistics

El Nacional (Venezuela)

New York Times

Oil & Gas Journal

OPEC Annual Statistical Bulletin

OPEC Bulletin
OPEC Review
Wall Street Journal

WEB SITES

BBC News (news.bbc.co.uk)
Energy Information Administration (www.eia.doe.gov)

Index

About the Authors

TOYIN FALOLA is the Frances Higginbothom Nalle Centennial Professor of History at the University of Texas at Austin. An expert on African history and international politics, he is the author of numerous books, including his memoir, *A Mouth Sweeter Than Salt* (2004) and *The Power of African Cultures* (2003).

ANN GENOVA is a doctoral candidate at the University of Texas at Austin. Her publications include "Oil in Nigeria" in *History in Africa.*